Video Surveillance and Social Control in a Comparative Perspective

Routledge Studies in Science, Technology and Society

1. **Science and the Media**
   Alternative Routes in Scientific Communication
   Massimiano Bucchi

2. **Animals, Disease and Human Society**
   Human-Animal Relations and the Rise of Veterinary Medicine
   Joanna Swabe

3. **Transnational Environmental Policy**
   The Ozone Layer
   Reiner Grundmann

4. **Biology and Political Science**
   Robert H Blank and Samuel M. Hines, Jr.

5. **Technoculture and Critical Theory**
   In the Service of the Machine?
   Simon Cooper

6. **Biomedicine as Culture**
   Instrumental Practices, Technoscientific Knowledge, and New Modes of Life
   Edited by Regula Valérie Burri and Joseph Dumit

7. **Journalism, Science and Society**
   Science Communication between News and Public Relations
   Edited by Martin W. Bauer and Massimiano Bucchi

8. **Science Images and Popular Images of Science**
   Edited by Bernd Hüppauf and Peter Weingart

9. **Wind Power and Power Politics**
   International Perspectives
   Edited by Peter A. Strachan, David Lal, and David Toke

10. **Global Public Health Vigilance**
    Creating a World on Alert
    Lorna Weir and Eric Mykhalovskiy

11. **Rethinking Disability**
    Bodies, Senses, and Things
    Michael Schillmeier

12. **Biometrics**
    Bodies, Technologies, Biopolitics
    Joseph Pugliese

13. **Wired and Mobilizing**
    Social Movements, New Technology, and Electoral Politics
    Victoria Carty

14. **The Politics of Bioethics**
    Alan Petersen

15. **The Culture of Science**
    How the Public Relates to Science Across the Globe
    Edited by Martin W. Bauer, Rajesh Shukla and Nick Allum

16 **Internet and Surveillance**
The Challenges of Web 2.0 and Social Media
Edited by Christian Fuchs, Kees Boersma, Anders Albrechtslund, and Marisol Sandoval

17 **The Good Life in a Technological Age**
Edited by Philip Brey, Adam Briggle and Edward Spence

18 **The Social Life of Nanotechnology**
Edited by Barbara Herr Harthorn and John W. Mohr

19 **Video Surveillance and Social Control in a Comparative Perspective**
Edited by Fredrika Björklund and Ola Svenonius

Video Surveillance and Social Control in a Comparative Perspective

Edited by
Fredrika Björklund and Ola Svenonius

NEW YORK   LONDON

First published 2013
by Routledge
711 Third Avenue, New York, NY 10017

Simultaneously published in the UK
by Routledge
2 Park Square, Milton Park, Abingdon, Oxfordshire OX14 4RN

First issued in paperback 2014

*Routledge is an imprint of the Taylor and Francis Group,
an informa business*

© 2013 Taylor & Francis

The right of the editors to be identified as the authors of the editorial material, and of the authors for their individual chapters, has been asserted in accordance with sections 77 and 78 of the Copyright, Designs and Patents Act 1988.

All rights reserved. No part of this book may be reprinted or reproduced or utilised in any form or by any electronic, mechanical, or other means, now known or hereafter invented, including photocopying and recording, or in any information storage or retrieval system, without permission in writing from the publishers.

Trademark Notice: Product or corporate names may be trademarks or registered trademarks, and are used only for identification and explanation without intent to infringe.

Library of Congress Cataloging-in-Publication Data
 Video surveillance and social control in a comparative perspective / edited by Fredrika Björklund and Ola Svenonius.
  p. cm. — (Routledge studies in science, technology and society ; 19)
 Includes bibliographical references and index.
 1. Video surveillance—Social aspects. 2. Social control.
 3. Privacy, Right of. I. Björklund, Fredrika. II. Svenonius, Ola.
 TK6680.3.V52 2012
 303.3'3—dc23
 2012017973

ISBN 978-0-415-62860-0 (hbk)
ISBN 978-1-138-92062-0 (pbk)
ISBN 978-0-203-08290-4 (ebk)

Typeset in Sabon
by IBT Global.

Contents

List of Figures	ix
List of Tables	xi
Acknowledgments	xiii

1 Video Surveillance in Theory and as Institutional Practice 1
   FREDRIKA BJÖRKLUND AND OLA SVENONIUS

PART I
Comparative Studies

2 Modernisation, Balancing Interests, and Citizens' Rights: Public Video Surveillance in Poland, Germany, and Sweden 19
   FREDRIKA BJÖRKLUND

3 Video Surveillance in a Historical Perspective 69
   OLA SVENONIUS

4 The Protection of Privacy in the Context of Video Surveillance: Towards a European Model? 97
   PATRICIA JONASON

PART II
Case Studies

5 Video Surveillance and the Question of Trust 131
   WOJCIECH SZRUBKA

6	How Effective is the Public Video Surveillance System in Warsaw? PAWEŁ WASZKIEWICZ	153
7	From Privacy Protection towards Affirmative Regulation: The Politics of Police Surveillance in Germany ERIC TÖPFER	171
8	The Pressure of the Practice: Swedish Public Surveillance in an Institutional Perspective ELFAR LOFTSSON	189

Contributors   213
Index   215

Figures

8.1 The concept of institution and its formative factors. 192
8.2 The structure of a practice-driven institution. 208

Tables

2.1	Video Surveillance as Discourse, Discursive Practice, and Social Regime	58
5.1	Interpersonal Trust I	150
5.2	Interpersonal Trust II	150
6.1	Resident Victimisation Rates in the Research Area During 12 Months Preceeding the Survey	161
6.2	Resident Victimisation Rates in the Research Area During 12 Months Preceeding the Survey	161
6.3	Residents of Four Areas within Warsaw Borders Feeling Safe in 2006 and 2007	164
6.4	Residents of Four Areas Feeling Safe in Their District in 2006 and 2007	165
6.5	Residents of Four Areas Feeling Safe in Their Yard and Block in 2006 and 2007	165
6.6	Residents of Four Areas Feeling Safe from Dusk until Dawn in 2006 and 2007	166

Acknowledgments

This book documents the project Balancing Integrity and Legal Security—A Comparison of Popular Surveillance in Germany, Sweden, and Poland, which was carried out at Södertörn University, Stockholm, 2008–2011. Thanks to the Foundation for Baltic and East European Studies for financing this project. Among several accommodating and helpful people who have contributed to making this book possible, we owe Dr. Michał Bron special thanks for his competent assistance with the interviews and field studies in Poland.

1 Video Surveillance in Theory and as Institutional Practice

Fredrika Björklund and Ola Svenonius

INTRODUCTION

One defining characteristic of the last twenty years is without a doubt the emergence of information and communication technologies (ICTs) as a central aspect in all areas of social life. Since the 1960s, and most notably after 1990 and the fall of the Iron Curtain, technological development has increased exponentially. Whereas the increasing dependence on, and belief in, technology's potential for making our lives easier is generally a positive feature of late capitalism, there is also reason to be very wary of how ICTs are deployed. This regards specifically the ever-increasing availability of relatively inexpensive ICTs developed for purposes of *surveillance* and *social control*. In this volume, the type of equipment that we are especially interested in is cameras, and the surrounding, attaching, and enhancing technologies that improve their functionality. But surveillance is much more than the electrified computer circuit linked to a lens and a digital receiver. What is interesting about surveillance practices are their motivations, practice, and institutional 'embeddedness'.

Surveillance in general, and video surveillance in particular, can be understood in terms of the indirect social control it facilitates, its central position in late modernity, and its function as a replacement for the direct social control that existed prior to the development of inexpensive ICTs. The discourse on surveillance in the social and human sciences owes much to the work of Foucault and Deleuze who, from an early stage, influenced critical scholars to take a very sceptical view of social control, as a perpetual machinery with no end to its appetite for increased discipline/control (Fiske 1998; Gandy 1993; Lyon 2006; Mathiesen 1989; Norris 2003). The possibility that surveillance technologies could be used for ill by authoritarian regimes was always there, but even in a democracy, freedom could be a *technology of power* as explored in Foucault's and Deleuze's work. During the 1970s and '80s, many Western European countries created strict institutional restrictions to prevent any trend towards the type of authoritarianism that existed in communist Eastern Europe, and which had been vividly described by George Orwell in *1984*. However, in the decades after

1989–90, as the old geopolitical distinction between East and West lost much of its meaning, and surveillance technology became more available, governments started to relax the strict regulations that were once in place. At the same time, new ideas concerning security have gained ground across Europe, which promote video surveillance as an effective means of crime prevention and to increase perceptions of security (Svenonius 2011). As the memory of Communism fades, new technological forms of social control have become dominant where before there was fear of authoritarianism. This book seeks to understand how that change came to be and how surveillance practices are institutionalised today.

This book reports the results of a research project on video surveillance, the West-East Video Surveillance Project,[1] carried out at Södertörn University in Stockholm between 2008 and 2010. Whereas policy diffusion throughout Europe has occurred to a large extent in the area of data protection, throughout the European Union (EU) large differences can still be observed in the area of video surveillance (Bennett and Raab 2006; Löfgren and Webster 2009; Norris 2012; Webster 2004). Video surveillance is therefore a worthwhile study subject because it is (still) mainly a national policy issue. The political nature of the surveillance problematic is closely related to the sensitive issues of political rights and legitimacy. It would probably be a difficult task for the EU to unite member states like Germany and Poland around a common solution regarding cameras, because, as we shall see herein, these countries have radically different approaches to video surveillance; even despite the fact that police forces in both Poland and Germany all advocate the effectiveness of video surveillance in combatting crime (see Eric Töpfer's and Paweł Waszkiewisc's contributions in this volume).

However, against this positive discourse stands empirical research, primarily the evaluations made during the first decade of the new millennium. Welsh and Farrington (2007), for example, show that video surveillance may reduce crime only to a small extent, and only when other preventive measures are also taken. Their work is hitherto the most extensive meta-evaluation of video surveillance, and the authors reach the conclusion that the marginal effectiveness is even more true for surveillance systems outside the United Kingdom, possibly due to the lack of other security measures used in parallel to video surveillance (Welsh and Farrington 2007: 57). This national specificity is one of the reasons that makes the institutionalisation of video surveillance such a strange phenomenon. The very different approaches to video surveillance that exist in the EU therefore constitute a valid field of study, and a rich environment for national comparison. The book is guided by certain questions: *How is video surveillance institutionalised in different national contexts? What are the differences in legal structures, political ambitions, and deployment practices?* Besides the tentative results from the Urbaneye Project (Hempel and Töpfer 2004), and the edited volume on video surveillance by Doyle, Lippert, and Lyon (2012), international comparative studies designed to answer this question have been rare.

This book concerns the institutionalisation of video surveillance, but it also addresses some other challenges facing the current surveillance literature, such as the dominance of the Anglo-Saxon perspective, and the lack of comparative studies, particularly between Eastern and Western Europe. In general, studies on video surveillance seem to remain predominantly national and rarely venture across borders. But since the UK and United States have become the forerunners in video surveillance applications and research, we have learnt too little about countries beyond the Anglo-Saxon cultural sphere. Despite the increasing range of empirical knowledge in the surveillance literature, non-Anglo-Saxon contexts are under-represented in the surveillance literature, and this becomes even more evident in the case of Eastern Europe, where empirical studies are still almost non-existent. One possible problem with the Anglo-Saxon dominance in the field is that culturally specific understandings of central concepts and societal problematics always risk being taken prima facie and becoming internalised as 'normal'. These conditions may have different meanings and are sometimes controversial where different notions of authority and statehood dominate.

In actuality, the problem is more fundamental than the issue of Anglo-Saxon dominance. Empirically more important is, as noted above, the *general* lack of comparative work on video surveillance. One reason for this is that video surveillance, although a central pillar of *urban* surveillance (Coaffee et al. 2009; Graham 2001) seems to have gone out of fashion. Today, researchers either focus on other types of technology or have shifted from a technologically determined focus to more general thematics such as governmentality, securitisation, or surveillance and 'mega-events'. It has also been said that the lack of a coherent research programme on surveillance has hindered the emergence of a proper knowledge base that can spark the advancement of comparative research. Marx's now famous description of the emerging genre of surveillance studies as "players without a field" (2007) is not without merit.

This situation motivated the research carried out at Södertörn University between 2008 and 2010. The aim of the West-East Video Surveillance Project was to deliver a substantial contribution to the empirical knowledge on surveillance in different countries, one that would allow us to grasp the contextual factors but at the same time broaden the approach using theories from the disciplines of sociology, law, and political science. In this book, we first and foremost let the empirical world guide the theory. The present volume is a collection of research reports that focus on video surveillance, its institutions, and social control in three countries: Germany, Poland, and Sweden. It gathers seven individual contributions that collectively provide comparative, descriptive, explanatory, and interpretive accounts of regulation, discourse, institutionalisation, implementation, and implications of video surveillance in all three countries, at both local and national levels.

In this introduction, we present the context of this book. First, we set out to discuss the issue of surveillance—that is, we situate the book in the relevant theoretical context. In short, we focus on surveillance as the institutionalisation of technologically mediated social control. This calls for a clarification of what we mean by *social control* and *institution*. Second, we discuss the implications of the comparative perspective as well as some methodological issues. Finally, we briefly introduce the contributions to the volume.

SURVEILLANCE AS A THEORETICAL CONCEPT

To social scientists, surveillance is interesting as a *practice* rather than as a *technology*. While many researchers in criminology and social science generally use surveillance as a prima facie concept—something empirically observable to all upon observation, much like a chair—the field known as *surveillance studies* approaches the concept analytically. It seeks to disentangle the characteristics of surveillance, to understand the ambiguities of different conceptualisations, and to bring out the implicit assumptions that are commonly not addressed outside this relatively young field. The field of surveillance studies was formed around the distinction between 'old' and 'new' surveillance, and around the emerging theories of 'disciplinary societies' (Foucault 1977), 'total surveillance societies' (Rule 1973), and 'maximum security societies' (Marx 1988).[2] There was, from the beginning, a tendency towards a technological focus—as in research related to closed-circuit television (CCTV)—and social constructionism in the wake of Foucault's influence from the 1970s onwards. The data protection literature, which had constituted a separate field of research and activism, merged to a certain degree with sociological theory towards the end of the 1980s, and became known as *surveillance studies*.

One of the problems with surveillance research since its inception has been its propensity towards over-theorising the object of analysis. The field of surveillance studies has seen extensive writing on discipline, liberty, and cultures of control, but often lacks a strong empirical base or, as Zureik puts it, "heavy on theorising and light on empirical research" (2007: 114). In our view, useful theories on surveillance are empirically manageable, give an enlightening account of the phenomenon, or are practice-oriented, focusing on surveillance as the purpose of data collection. Marx's distinction between old and new surveillance (2002), Haggerty and Ericson's application of Deleuze and Guattari's assemblage theory (2000), and Norris and Armstrong's empirically based formulation of the political problem with surveillance (1999) are all examples of influential works that do not over-theorise the topic and in different ways allow us to understand the phenomenon of surveillance in a better way.

Conceptually, we can see two main trends in the surveillance studies of how to understand surveillance as such—that is, as opposed to surveillant practices. On the on hand, there are sociological theories with a Foucauldian and sometimes Marxist heritage which highlight social control and disciplinary power, and collective autonomy; one such example is, "Surveillance involves the collection and analysis of information about populations in order to govern their activities" (Haggerty and Ericson 2006: 3). Other ways of understanding the concept exist, such as Fiske's, which conceives of surveillance as a "technology of whiteness that racially zones city space by drawing lines that Blacks cannot cross and whites cannot see" (1998: 69); and Lyon's term 'social sorting' (2003, 2007a) is a way to pinpoint surveillance as a mechanism aimed at seamlessly separating and categorising groups. The commonality between Haggerty and Ericson's, Fiske's, and Lyon's concepts of surveillance is clear, and partially lies in their intellectual debt to Foucault, who discussed surveillance as a technique of domination that acts both on a granular level within each subject and as a form of power (Foucault 1977, 2007). The intentionality, the analytical focus on populations, and the power/knowledge nexus are all quite specific aspects of these authors' ways of understanding the topic.

A different way of conceptualising surveillance in more traditionally liberal terms would be to view it as an unlawful breach of the social contract when the monitored subject does not consent to being monitored. Surveillance from this data protection perspective can be defined as 'the purposive collection of personal data'. This way of understanding surveillance is perhaps more suited to legal analyses, such as Patricia Jonason's chapter in this volume, because it highlights the issue of privacy in relation to surveillance, which is the normative backbone in the analyses. The idea of privacy used to be quite central to surveillance studies, but it has more or less been declared a 'lost fight'. Today, many scholars rather speak of 'liberties' in general than of 'privacy' (Bigo et al. 2010; Coaffee 2009). One reason for the decline in the analytical value of privacy might be because it has been quite thoroughly regulated, at least in Europe, and therefore depoliticised to an extent. As Eric Töpfer argues in this volume, we should not let the idea of privacy go but instead follow Regan (1995), who speaks of privacy as a *collective* value. The distinction between surveillance as *social control* and surveillance as *personal data collection* is arguably a worthwhile one, and a defining characteristic of surveillance studies.

In this book, surveillance exists somewhere between the positions discussed above. It is a technique of data gathering that performs the function of social control, irrespective of whether we are speaking about consumers, citizens, passengers, guests, or patients;[3] but it is also a way of collecting personal data. Common to all contributions is the centrality of institutions (governmental bodies, law enforcement, and certain cultural practices). We focus on *institutional practice*, which arguably should be at the core of every theory on surveillance, whether Foucauldian, post-Marxist, or liberal. Institutions are the sites where cultural and historical narratives are reproduced, where norms are created and enforced, and where collectives exert control over individuals. This is a central understanding of surveillance that is remarkably absent in the surveillance studies literature and is something we try to remedy with this volume. Our argument is that in social science, surveillance is most interesting as an institutional practice—that is, as a collective endeavour, whether it be on Web 2.0, in urban security governance, in public discourse, in international police cooperation, or in regulation activities. Outside of the institutional focus, the distinction of old and new surveillance, the surveillance assemblage, and the political consequences of extensive CCTV are stripped of the power dimension and therefore risk losing much of their interesting complexity.

Institutional perspectives on surveillance can, however, take very different forms, which the contributions in this volume display with great clarity. Elfar Loftsson and Wojciech Szrubka stay close to the political science discourse on 'the new institutionalism' and seek to apply and develop this theoretical perspective. Ola Svenonius writes about surveillance practices as outcomes of institutional settings, while Fredrika Björklund focuses on discourses and their institutional effects. Patricia Jonason, Eric Töpfer, and Paweł Waszkiewicz, finally, focus on institutional practice in terms of regulation and law enforcement. Thus, even though all are concerned with issues of institutional power, each of the chapters focuses on different aspects of surveillance practices and highlights different institutional practices in each context. We share the aim of shedding light on the structural conditions for power and revealing aspects of this area that are not obvious to everyone. The social control approach to video surveillance opens up a critical agenda but is reflected in this book more as an approach to setting the questions to be asked than as a programme for policy improvements. At the end of this introduction, the chapters will be presented in more detail.

The present volume makes a contribution to the social sciences in general, not the least in terms of the new empirical data presented. The institutional theoretical framework places the studies within the disciplinary framework of political science. In fact, we are perplexed by the sparse interest that political scientists have shown in the surveillance problematic, and we certainly want to remedy this. Surveillance studies as an academic field—with a strong connection to sociological theory—has guided the discussion on social control. Both these fields are expected to benefit from the new empirical findings that we present. But since parts of the empirical material concern Eastern Europe, we also aspire to make some contribution to the fields of Eastern European or post-communist studies. In the next section we more thoroughly discuss issues of comparative research and gaps in the literature before moving on to a brief presentation of the book's chapters.

SURVEILLANCE IN A COMPARATIVE PERSPECTIVE

This study involves three European countries: Germany, Sweden and Poland. It was guided by the idea that each society's historical trajectory has differences that are manifest in the methods of video surveillance regulation and deployment. What puzzled us was that technological surveillance, although expanding all over the world, seems to fulfil different functions in different countries' specific narratives and modes of regulation. Arguments for monitoring in a democracy certainly have to be more embedded in other contextual considerations than in an autocracy. In a democratic society, legitimacy is (ideally) constructed around popular consent, such that the introduction of coercive technologies is precluded by public debate. In an ideal case, democratic legitimacy for surveillance has two aspects: accountability and effectiveness. Accountability guarantees that surveillance practices are not allowed to 'creep', and thus threaten democratic values. Effectiveness implies that there is a societal gain to be expected from surveillance. It is not controversial to assume that modes of legitimisation, explanation, accountability, and argumentation for the effectiveness of CCTV differ between societies. The general public has to be convinced that cameras have safety benefits and/or it has to be persuaded that video surveillance is a sufficient and necessary means of combatting crime. In a non-democratic state, these legitimacy claims are superfluous, of course, but in the societies of democratic Europe there is reason to study how different modes of regulation, legitimisation, and administration of surveillance functions. These are the types of differences that this book sets out to explore.

This volume is intended to fill the gap in the literature on surveillance as regards comparative perspectives and the legacy of communism in contemporary democratic societies. We thus set out from the assumption that there are narrative and contextual differences between Western and Eastern Europe in relation to monitoring, and this should be particularly true for the selection of cases in this study. Arguably, the 'surveillance society' is integrated in national discourses in various ways, some of which are discussed in the chapters herein. It should be stressed that we are not primarily interested in finding causal explanations for differences by comparing countries. We are, rather, looking for different national traits that help us to get a deeper understanding of the phenomena of video surveillance and the differences in social organisation that they represent. From the start, we were convinced that such differences existed and that they were important, but we had also noticed the lack of relevant academic research on these issues. Beyond the Urbaneye Project, which was carried out in the years 2001–2004 and took a snapshot of video surveillance in eight European countries, comparative research in this area was virtually non-existent.

Germany, Poland, and Sweden are located in the same geographical region, and they are all EU member states, but the differences in political history are immense. Consider, for example, a country such as Poland, where—although it is a steady member of the EU and other Western collaborations—democracy is still relatively new. However, its familiarity with an authoritarian political system and a repressive secret police does not seem to have hampered the introduction of video surveillance. On the contrary, Poland is far ahead of most other EU member states when it comes to extensive video surveillance systems. In which national narrative is this activity embedded? Do remnants of the former communist regime affect the present institutional setting? Next, consider Sweden, with its firm democratic tradition to fall back on. The need for monitoring was constructed so that supervision of the public was compatible with the democratic values on which the Swedish state was based. What reasoning and what kind of institutions were needed to solve this problem? What does this rationale look like in comparison with the Polish 'psyche'? Our third case, Germany, definitely has a different political history than both Sweden and Poland. The reunified Germany has experiences from both sides of Europe's recent history. But more important in this context is the tradition stressing the *Rechtsstaat* aspects of democracy. The German institutional setting clearly visualises the complex contradiction between civil rights and social control through surveillance. Germany was, for this reason, deemed to be of high interest not only to researchers in the respective countries themselves but to surveillance studies and related research areas in general.

The comparative ambitions of this book are, in a way, unique. As discussed above, surveillance studies has traditionally been mainly an Anglo-Saxon research field (cf. Murakami Wood 2009). The main empirical works have been authored by British and North American authors, and nationals from these countries still dominate the discipline. The English-speaking countries of the world are over-represented as cases in the literature, although a considerable internationalisation has been taking place in recent years. The Anglo-Saxon dominance in the field, as elsewhere, is related to the richness of Britain to studies of CCTV. However, as Goold reminds us, when empirical data stem from only a few cases, there is a risk that the contingencies of those cases are blurred. Goold writes, "Yet has the rise of surveillance been as uniform as the experiences of the UK and the USA might suggest? Have developments in Europe mirrored those in the English-speaking West? Although surveillance studies is an international discipline, there has been a tendency for those in the field to examine issues surrounding surveillance through a particularly Anglo-American lens" (Goold 2009: 116).

In other words, we might run the risk of generalising policy-making experiences from the UK to other countries, despite the fact that these societies belong to totally different policy regimes in key areas (Crawford 2009). Even more important are the conceptual issues that arise out of the Anglo-Saxon dominance of surveillance studies and related fields such as criminology, sociology, and political science. These conceptual issues urge us to do studies from other parts of the world, particularly comparative studies. Below we address issues of universality, which make comparative work more complicated than it first looks.

When researching surveillance, sensitivity to context is of particular importance, precisely because surveillance is so intimately related to statehood, citizenship, and political liberties. The countries where surveillance studies has been most prevalent—the United Kingdom, the United States, Canada, and to some extent (formerly West) Germany—are also countries characterised by comparatively strong liberal sentiments, where there is a strong tradition of the right to be left alone from state interference. The British case is a clear example: with its radical liberalisation during the years of Prime Minister Margaret Thatcher, a major dislocation occurred in British society. Video surveillance came as part and parcel with the general adjustment of the public sector to the market and management practices of the 1990s. It represented a forceful response to demands for clean and peaceful urban areas and at the same time it appeared as an individualised form of social control. To some extent, crime prevention via the use of cameras has replaced the traditional ambitions of social prevention. That this would generate interest from civil society organisations and the academic community is not surprising.

In *Sweden*, on the contrary, the academic community has been fairly ignorant of the issue of privacy. In fact, it was a symbol and tool of conservative and liberal forces against the development of the welfare state, and therefore antagonistic to the social democratic enterprise *per se*. Even so, Sweden introduced the first data-protection act in the world. Accordingly, there are certain tensions in Swedish politics regarding surveillance that cannot be understood without paying careful attention to specific national conditions, and Fredrika Björklund details some of these aspects in her chapter in this volume. In other countries, such as Poland, conducting surveillance studies would have meant imprisonment until fairly recently. Today, surprisingly few works on surveillance in Eastern Europe have reached an international audience. In Poland, one of the countries analysed here, Waszkiewicz's dissertation (2011) is the first proper study on video surveillance. Needless to say, Polish society is immensely interesting as a study object from a surveillance point of view, and is subject to quite different mechanisms than that of any of the Anglo-Saxon countries. Maria Łoś has written extensively on the privatisation of the Polish 'police state', and shows how Poland's secret service networks have managed to use the velvet revolutions to their own advantage (see, e.g., Łoś and Zybertowicz 2000, Łoś 2002). The *German* case, finally, is also very intricate. The regime change in the former East Germany was not very easy for large segments of the population, and there is a growing body of research dealing with identity construction in the transition from authoritarian regime to democracy (Gerber 2011). As Kunz (2005) and Hannah (2008, 2010) show, the legacy of the fight against terrorism in the 1970s forms an important normative point of departure for German scholars critical of interior security (*innere Sicherheit*). The 1970s also witnessed a major dislocation in German political consciousness that cannot be understood without reference to the historical experiences of fascism during the Third Reich. Thus, in all three cases in this study, we see historical contingencies that (unsurprisingly) deviate from the British, US-American, Canadian, Irish, and Australian experiences.

All three countries that are included in this study are liberal democracies today, but as stated above, both Poland and Germany have recent experiences of non-democratic regimes. Here we think that there are very fertile grounds for comparative study, as the contributions by Fredrika Björklund, Ola Svenonius, and Patricia Jonason show. In addition, the case studies present deeper analyses of the individual countries, which yield more in-depth knowledge of the processes at work. This volume attempts to expand both knowledge about video surveillance in Europe and the range of perspectives beyond those that hitherto have dominated the field. Surveillance studies are in dire need of comparative studies outside Western Europe and North America—that is, from countries where cultural, social, and political factors cannot be discarded by claiming *ceteris paribus* (Löfgren and Webster 2009).

Still, there is reason to focus attention on some methodological issues that must be considered in comparative works of this kind. To some extent, these issues have to do with language and translation, but linguistic expressions also reflect the conceptual differences that are the focus of the study and which must be brought to light. We use the Swedish word for privacy—*integritet*—and juxtapose its translation into English. We figured that the aim was to compare the different positions of each country in terms of the balance between security and privacy on the issue of surveillance. Now, one aspect of the project was to investigate exactly those semantic factors, and examine their role in how discourses on security and privacy are constructed, rather than simply assuming their universal sameness. In that process, the 'balance' between privacy and security became the object of analysis not only from a discursive perspective but also from institutionalist and legal ones. The first part of this book deals extensively with this aspect of the surveillance thematic.

A notion frequently used in the book is a conception of video surveillance as 'public surveillance'. The notion of public surveillance can be contested, because of its vagueness (why *public*, and what does it stand for?). What we really mean is the anthropological notion of everyday, mundane activities, the monitoring of which is what video surveillance has become. It is so common today for any café, barbershop, university, bank, underwear store, or public transport system to use this technology extensively. There are few other surveillance technologies that are so mature and normalised today as are cameras. There is still awareness about, for example, privacy infringements on the Internet, but only a few people consider cameras to be controversial anymore. Rather, the camera has taken on a symbolic position in the practice of, and opposition to, surveillance in general, but the reasons why people used to oppose the use of cameras have largely fallen into obscurity. Thus, we wanted to analyse public video surveillance because we believe that such surveillance, more than any other kind, is directed

Video Surveillance in Theory and as Institutional Practice 11

towards the average citizen in the street. The phenomenon of public video surveillance still touches upon questions that have to do with what the state wants from its citizens. Still, it should be remembered that when we use the word *public* it is not completely equivalent to the Swedish legal terminology where it used to serve as a prefix before 'video surveillance' (*allmän kameraövervakning*). Languages often don't allow for direct translations of such subtleties, and we have to be aware of the Anglo-Saxon linguistic lens through which we have to analyse our material.

THE CONTRIBUTIONS IN THIS VOLUME

Although the majority of the authors in this volume are political scientists, legal and sociological approaches are also represented. This disciplinary range is of great value for enriching this study. One other advantage of this study is the complementary political science perspectives it provides on a study field dominated by sociologists and, when it comes to related issues of human and civil rights, scholars of law. Below is a short overview of each chapter in the volume.

The basic premise behind the various practices discussed in the book is surveillance as *institutionalised social control*. There are no strict definitions of 'institution' that tie the contributions together. Rather, each contributor, with the institutional perspective in mind, approaches the field of study in accordance with their particular methodological interest. Some of the chapters deal with institutions explicitly and in a formalised way, others more implicitly. However, what *is* common to the different approaches is the rejection of a narrow definition of institutions in favour of a definition that includes values, rules, and practices. But the emphasis is different in each of the individual chapters, which are sorted into two parts: *comparative* studies and *case* studies. The comparative studies involve all three countries, are tied together by the institutional perspective, and give *legal*, *historical*, and *discursive* perspectives on video surveillance. Put together, the three chapters give a rich picture of national surveillance policies and the differences and similarities between them.

The first chapter, "Modernisation, Balancing Interests, and Citizens' Rights: Public Video Surveillance in Poland, Germany, and Sweden", by Fredrika Björklund, concerns monitoring as discourse. Significant elements of national discourses are explored and contrasted. Poland, Sweden, and Germany represent three different discursive approaches to video surveillance that are also closely connected to particular institutional settings. The Polish discourse on surveillance is characterised by the idea of modernisation and the adoption of a Western European standard of privacy rights. In the Swedish case, the conditions for monitoring are set by utilitarian principles, which are represented in the established idea of a balance between interests that can be traced back to a strong welfare-state tradition. Unlike the other two countries, video surveillance in Germany is seriously contested, and the objections concern its legality. The German discourse originates from historical threats against the rule of law.

In the second chapter, "Video Surveillance in a Historical Perspective", Ola Svenonius discusses the historical developments that preceded, and still to an extent define, video surveillance practices in all three societies today. Through a focus on political struggles over the framing of the surveillance and privacy problematic, Svenonius provides an account that shows how law enforcement and data protection institutions have developed from roughly 1970 until today. Through both historical and cross-country comparison, Svenonius discusses the possibilities for convergence and for a critical debate on video surveillance. Whereas today's situation, from a privacy perspective, may seem most problematic in Poland, similar tendencies are also observed in Sweden and, to a lesser extent, in Germany. Video surveillance is a most certainly a technology on the move, as new schemes are conceived in an increasingly positive discourse on video surveillance as an invaluable tool in the fight against crime.

In the third chapter of the volume, "The Protection of Privacy in the Context of Video Surveillance: Towards a European Model?", by Patricia Jonason, the basic legal instruments affecting video surveillance practices in Germany, Poland, and Sweden are presented and discussed. Relevant legal documents are examined, including national acts and constitutions, as well as European instruments. The legal situations in these countries make for interesting comparisons that deepen the understanding of the different issues that are discussed in the following chapters. Although often neglected by social science scholars, legal documents certainly belong to the institutional setting of surveillance practices.

The second part of the volume consists of four case studies, which each discuss nation-specific issues in one of the three countries. Still, all chapters include matters of general interest to the field. The first study explores the way attitudes to surveillance connect to interpersonal trust in a society—a highly relevant topic not only in former communist states such as Poland but also elsewhere. The second chapter concerns the efficiency of technical supervision, which is a subject that has been discussed by many scholars and practitioners. The third chapter is about the expansion of surveillance as an outcome of political and institutional power struggles, and this is a perspective that is strongly relevant in times when governments are supposed to be addressing conflicting values such as freedom and security. The fourth and final chapter takes a consistent institutional approach to video surveillance and investigates the normative aspects in contrast to the practice-driven implementation of law.

"Video Surveillance and the Question of Trust", by Wojciech Szrubka, explores the popularity of surveillance in Poland as an outcome of a general lack of interpersonal trust. The popularity of video surveillance among the public is puzzling in many countries, but the Polish case gives the dilemma an extra dimension because of its political history, where the former communist regime used secret supervision as a technique of social control. Certainly one would assume that, in the modern democratic Polish state, technical surveillance would be very much disliked by the citizenry; but this is not the case. Using a rational choice perspective on social trust, the chapter raises the question of whether this attitude could be seen as a product of destroyed social trust engendered during communist times.

The next chapter is also about video surveillance in Poland, but from a different angle. In "How Effective is the Public Video Surveillance System in Warsaw?", Paweł Waszkiewicz studies the crime-preventing effects of video surveillance in the Polish capital. Is video surveillance an efficient way of combatting crime and increasing the feeling of security among residents? The study is constructed as a quasiexperiment, comparing two experimental areas with two control areas. The results in no way contradict international studies asking similar questions, but this investigation gives an extra dimension to the issue because of its in-depth perspective on the topic and the interesting design of the study. In this book we present one of the first academic studies of these matters in Poland.

Eric Töpfer, in his contribution, "From Privacy Protection towards Affirmative Regulation: The Politics of Police Surveillance in Germany", analyses the expansion of surveillance in a particularly German context of a legally established right to informational self-determination. Despite this unique understanding of privacy, established by the Federal Constitutional Court in 1983, strong advocacy coalitions have managed to turn the debate in their preferred direction of continued expansion. The process of progressively curtailed citizens rights, from the modernisation of policing in Germany in the 1970s to the current legal regulations, are thoroughly discussed and problematised.

In the last contribution, Elfar Loftsson's "Pressure of the Practice: Swedish Public Surveillance in an Institutional Perspective", video surveillance in Sweden is discussed within the framework of new institutionalism. The aim of this chapter is first to survey the concept of institution for a coherent model of analysis and second to use the model tentatively to frame, describe, and analyse video surveillance institutions in Sweden. In this chapter it is argued that due to rapid technological development, increased international cooperation in combatting crime, and internal political needs to demonstrate strength, the practices of institutions tend to shrink the importance of values, norms, and rules.

NOTES

1. The formal name of the project is *Balancing Integrity and Legal Security: A Comparison of Popular Surveillance in Germany, Sweden and Poland.*
2. See Lyon 2007b for a discussion of the roots of surveillance studies.

3. Social control signifies "those organised responses to crime, delinquency and allied forms of deviant and/or socially problematic behaviour which are actually conceived of as such, whether in the reactive sense (after the putative act has taken place of the actor has been identified) or in the proactive sense (to prevent the act)" (Cohen 1985: 3). Institutionalised social control is represented in government, the legal system, and more or less formally in the act of monitoring, etc.

BIBLIOGRAPHY

Adelman, Jonathan R. 1984. *Terror and Communist Politics: The Role of the Secret Police in Communist States*. Boulder, Colo.: Westview Press.
Bennett, Colin J., and Charles D. Raab. 2006. *The Governance of Privacy: Policy Instruments in Global Perspective*. 2nd ed. Cambridge, Mass.: MIT Press.
Bigo, Didier, Sergio Carrera, Elspeth Guild, and R. B. J. Walker, eds. 2010. *Europe's 21st Century Challenge: Delivering Liberty*. Farnham, England: Ashgate.
Bruce, Gary. 2003. "The Prelude to Nationwide Surveillance in East Germany Stasi Operations and Threat Perceptions, 1945–1953." *Journal of Cold War Studies* 5(2): 3–31.
Coaffee, Jon. 2009. *Terrorism, Risk and the Global City: Towards Urban Resilience*. Farnham, England: Ashgate.
Coaffee, Jon, David Murakami Wood, and Peter Rogers. 2009. *The Everyday Resilience of the City: How Cities Respond to Terrorism and Disaster*. Basingstoke, England: Palgrave Macmillan.
Cohen, Stanley. 1985. *Visions of Social Control: Crime, Punishment, and Classification*. Cambridge: Polity Press.
Crawford, Adam. 2009. "Situating Crime Prevention Policies in Comparative Perspective: Policy Travels, Transfer and Translation." In *Crime Prevention Policies in Comparative Perspective*, ed. Adam Crawford. Devon, England: Willan, 1–37.
Doyle, Aaron, Randy Lippert, and David Lyon, eds. 2012. *Eyes Everywhere: The Global Growth of Camera Surveillance*. London: Routledge.
Fiske, John. 1998. "Surveilling the City: Whiteness, the Black Man and Democratic Totalitarianism." *Theory, Culture, and Society* 15(2): 67–88.
Foucault, Michel. 1977. *Discipline and Punish. The Birth of the Prison*. New York: Vintage.
Foucault, Michel. 2007. *Security, Territory, Population. Lectures at the College de France, 1977–78*. Houndmills, Basingstoke, Hampshire: Palgrave Macmillan.
Gandy, Oscar. 1993. *The Panoptic Sort: Towards a Political Economy of Information*. Boulder, Colo.: Westview Press.
Gerber, Sofi. 2011. *Öst är Väst men Väst är bäst: Östtysk identitetsformering i det förenade Tyskland*. PhD. diss. Stockholm: Stockholm University.
Goold, Benjamin J. 2009. "Editorial: Making Sense of Surveillance in Europe." *European Journal of Criminology*. 6(2): 115–117.
Graham, Stephen. 2001. "CCTV: The Stealthy Emergence of a Fifth Utility?" *Planning Theory and Practice* 3(2): 237–41.
Haggerty, Kevin D., and Richard V. Ericson. 2000. "The Surveillant Assemblage." *British Journal of Sociology* 51(4): 605–22.
———. 2006. *The Politics of Surveillance and Visibility*. Toronto: University of Toronto Press.

Hannah, Matt. 2008. "Die umstrittene Konstruktion von Vertrauen und Misstrauen in der westdeutschen Volkszählungsboykottbewegung 1983." *Social Geography* 3: 11–21.

———. 2010. *Dark Territory in the Information Age: Learning from the West German Census Boycotts of the 1980s*. Farnham, England: Ashgate.

Hempel, Leon, and Eric Töpfer. 2004. *CCTV in Europe: Final Report*. Urbaneye Working Paper No. 15. Berlin: Centre for Technology and Society.

Kunz, Thomas. 2005. *Der Sicherheitsdiskurs: Die innere Sicherheitspolitik und ihre Kritik*. 1st ed. Bielefeld, Germany: Transcript-Verlag.

Löfgren, Karl, and C. William R. Webster. 2009. "Policy Innovation, Convergence and Divergence: Considering the Policy Transfer Regulating Privacy and Data Protection in Three European Countries." *Information Polity: The International Journal of Government and Democracy in the Information Age* 14(4): 279–98.

Łoś, Maria. 2002. "Post-communist Fear of Crime and the Commercialization of Security." *Theoretical Criminology* 6(2): 165–88.

Łoś, Maria, and Andrzej Zybertowicz. 2000. *Privatizing the Police-state: The Case of Poland*. Basingstoke, England: Macmillan.

Lyon, David. 2003. *Surveillance as Social Sorting: Privacy, Risk, and Digital Discrimination*. New York: Routledge.

———. 2006. *Theorizing Surveillance: The Panopticon and Beyond*. Portland, Ore.: Willan.

———. 2007a. "Surveillance, Security and Social Sorting." *International Criminal Justice Review* 17(3): 161–70.

———. 2007b. *Surveillance Studies: An Overview*. Cambridge: Polity Press.

Marquardt, Nadine. 2005. "Vom Deutscher Herbst und 11. September die Konstruktion nationaler Identität im Diskurs der inneren Sicherheit." Unpublished manuscript, University of Berlin.

Marx, Gary T. 1988. *Undercover: Police Surveillance in America*. Berkeley and Los Angeles: University of California Press.

———. 2002. "What's New about the "New Surveillance"? Classifying for Change and Continuity." *Surveillance and Society* 1(1): 9–29.

———. 2007. "Desperately Seeking Surveillance Studies: Players in Search of a Field." *Contemporary Sociology* 36(2): 125–30.

Mathiesen, Thomas. 1989. *Den dolda discipineringen: Essäer om politisk kontroll Makt och motmakt*. Göteborg, Sweden: Korpen.

Murakami Wood, David. 2009. "Situating Surveillance Studies: Sean Hier and Josh Greenberg's *The Surveillance Studies Reader*, and David Lyon's *Surveillance Studies*: An Overview." *Surveillance and Society* 6(1): 52–61.

Norris, Clive. 2003. "From Personal to Digital: CCTV, the Panopticon, and the Technological Mediation of Suspicion and Social Control." In *Surveillance as Social Sorting: Privacy, Risk and Digital Discrimination*, ed. David Lyon. London: Routledge, 249–81.

———. 2012. "There's no success like failure and failure's no success at all: some critical reflections on the global growth of CCTV surveillance." In *Eyes Everywhere: The Global Growth of Camera Surveillance*, ed. Aaron Doyle, Randy Lippert, and David Lyon. Oxon, New York: Routledge, 23–45.

Norris, Clive, and Gary Armstrong. 1999. *The Maximum Surveillance Society: The Rise of CCTV*. Oxford: Berg.

Norris, Clive, Mike McCahill, and David Wood. 2004. "Editorial. The Growth of CCTV: A Global Perspective on the International Diffusion of Video Surveillance in Publicly Accessible Space." *Surveillance and Society* 2(2–3): 110–35.

Regan, Priscilla M. 1995. *Legislating Privacy: Technology, Social Values, and Public Policy*. Chapel Hill: University of North Carolina Press.

Rule, James B. 1973. *Private Lives and Public Surveillance*. London: Allen Lane.
Svenonius, Ola. 2011. "Sensitising Urban Transport Security: Surveillance and Policing in Berlin, Stockholm, and Warsaw." PhD diss., Department of Political Science, Stockholm University.
Waszkiewicz, Paweł. 2011. *Wielki Brat Rok 2010. Systemy monitoringu wizyjnego—aspekty kryminalistyczne, kryminologiczne i prawne*. Warsaw: Wolters Kluwer Polska.
Webster, William R. 2004. "The Diffusion, Regulation and Governance of Closed-Circuit Television in the UK." *Surveillance and Society* 2(2–3): 230–50.
Welsh, Brandon C., and David P. Farrington. 2007. *Kameraövervakning och brottsprevention: En systematisk forskningsgenomgång*. Stockholm: Brottsförebyggande rådet (Brå).
Zureik, Elia. 2007. "Surveillance Studies: From Metaphors to Regulation to Subjectivity." *Contemporary Sociology*. 36(2): 112–115.

Part I
Comparative Studies

2 Modernisation, Balancing Interests, and Citizens' Rights
Public Video Surveillance in Poland, Germany, and Sweden

Fredrika Björklund

INTRODUCTION

Today, the practice of using specially installed video cameras for keeping the public under surveillance has become well-established in much of the world (Norris et al. 2004). In many places, video cameras have become an accepted feature both on streets and in indoor shopping galleries. At the same time, this kind of surveillance has been neglected by critics, who have tended to focus more on the Internet, biometrics, and other technologically innovative forms of surveillance. It is a fact, however, that video surveillance, when used to control physical public space, leaves a clear mark on the state-citizen relationship. The normalisation of video surveillance indicates a lasting change in this relationship. Much research remains to be done on the implications of this change. This chapter will use a comparative approach to increase our understanding of the discourses that emerge in the wake of such surveillance.

This study is motivated, in part, by our interest in video surveillance as an expression of a changed relationship between the state and citizen, with special attention to the state's authorisation to control its citizens. The state is not treated as a monolithic actor; we look, rather, at the state's representatives, in the form of the various public institutions and actors who use surveillance. Our research has also been framed by the hypothesis that the social implications of video surveillance vary from nation to nation, in accordance with each nation's historically conditioned attitudes towards the role of the citizen. In order to understand video surveillance as a phenomenon, and to grasp its underlying meaning, it must be placed in social and political context. Otherwise, video cameras are nothing more than an expression of people looking at each other. The national context is primary here, for video surveillance is covered by very few international political practices or systems of international norms. Depending on the case, the expansion of video surveillance as an instrument of control may have completely different import.

The three countries examined in this study are Sweden, Poland, and Germany. Although all are, today, members of the European Union, they cannot be said to share a common European history. Two of them, Germany and Poland, have relatively recently experienced totalitarian government. Does this have an impact on their surveillance practices today? Similarly, does it matter that Sweden has been spared a similar fate? A natural hypothesis would seem to be that countries with recent experiences of totalitarianism would be more cautious concerning all sorts of surveillance ambitions. As Kilian argues in regard to Germany: "For post-war Germans, sensitivity to the need to protect information came easily. Recent historical experience of totalitarian government . . . made Germany one of the first countries in the world to adopt privacy protection codes" (Kilian 2008: 80).

Szekely draws similar conclusions regarding Hungary, another European state whose history is marked by Europe's dark past (2008: 175). It seems reasonable to assume that there is a social logic at work here. At the same time, such an inference may be too simplistic—a suspicion confirmed by the example of Poland, which in the last decade has seen a powerful expansion of video surveillance. Here, we have a case of continuity; old forms of governance and attitudes seem to be reproduced, not rejected. On the other hand, one must not assume a direct correlation between modern Poland's video surveillance and the surveillance of which the communist regime was guilty. It is more likely that historical and national continuity operates not directly, but through indirect expression.

There are significant differences between public surveillance conducted in the former Eastern bloc and that witnessed today in the democratic world. Video surveillance in the three countries examined here is carried out largely by private actors, subject to varying degrees of legal regulation. Most often, private interests and companies take the initiative of establishing video surveillance. Public institutions and controlling authorities benefit from this surveillance and use it for public and governmental purposes—for example, to fight crime and terrorism. Public authorities such as the police likewise conduct varying degrees of surveillance, but the actual business of running surveillance operations is usually outsourced to various network entrepreneurs and private actors (Newman 2008: 4; Webster 2004). The changed state-citizen relationship of which video surveillance is an expression is also mediated through a general social development in which *government* is replaced by *governance*. This is a transnational trend. Nonetheless, the practice of video surveillance varies according to national institutions, attitudes, and norms. The discourse analysis approach used here is particularly well suited to contribute to our knowledge of how this process manifests itself.

Our study reveals three different discourses on video surveillance: surveillance as balance of interests, surveillance as modernisation, and the discourse on surveillance in relation to citizens' rights. Our categories are based on empirical evidence charting dominant attitudes towards surveillance in each country. In Sweden, the dominant discourse approaches surveillance in terms of a balance of interests. This discourse structures what is said and done. It is also tied to a particular institutional organisation. As a whole, the Swedish discourse creates an order that can be traced to the legacy of a well-established welfare state. Here, surveillance can be justified as an activity that harmonises with traditional state responsibilities. The Polish discourse, which links surveillance to modernisation, also serves a legitimatising function. In this case, the discourse identifies Poland as a modern, democratic, European country. Video surveillance in Germany, finally, is placed within a larger discourse concerning the rights of citizens and the legal regulation of citizenship, issues to which both supporters and opponents must relate. The German discourse is a reflection of the country's far-reaching and continuous social problematisation of a national legacy that includes both a strong rights tradition and the experience of totalitarianism.

These three discourses—balance of interests, modernisation, and citizens' rights—are, of course, not the only discourses of surveillance in existence. Another surveillance discourse concerns what is termed *the risk society*. This discourse is rooted in the idea that society is becoming increasingly dangerous to the individual and that, in consequence, risk management must be one of primary responsibilities of political authority. In an era of terrorism and increasing crime rates, the state must protect and provide security for its citizens. This is a relatively widespread discourse concerning (late) modern society (Luhmann 1993; Beck 1992). In this study, we treat this discourse as the basis upon which the three above-mentioned discourses depend. The perceptions of increasing threats to society and the individual are, we believe, a necessary precondition for the legitimatisation of video and other types of surveillance in a democratic society.

VIDEO SURVEILLANCE AS AN EXPRESSION OF A SOCIAL REGIME

At a general level, one may say that video surveillance as a phenomenon belongs to a late form of modernity, or, using Beck's expression, 'reflexive modernity', in which the traditional lines of political conflict have been blurred (1992). In late modern societies, video surveillance can remain relatively uncontroversial—that is, apolitical—and the exercise of power that it represents can be hidden from the majority of citizens. Video surveillance, like other forms of surveillance, is seen as a sort of neutral instrument in a society where political energy and the tension between political interests have been re-directed towards protecting the established order and minimising risk. The three cases studied here can be viewed as empirical illustrations of this. Video surveillance does not attract very much political interest, nor does it mobilise social movements. Here, as we shall see below, Germany may be an exception; among the three countries in this study, only Germany shows evidence of political activity in response to video surveillance.

The denial of the political as a dimension of conflict is characteristic of the modern liberal project (Mouffe 2005). According to this perspective, society is composed of individuals whose aggregated wills express a rational consensus which puts an end to conflicts of interest. This liberal (some would say reductionist) view of politics is, at present, dominant. It colours our perception of social issues and social relations in general. The image of an autonomous individual with free choice, existing outside the political, is the foundation of the last decades' re-negotiation of state and government power in the West. Particular aspects of late modern society seem to follow close on the heels of this de-politicisation. The expanding and perhaps totalitarian surveillance that Norris and Armstrong describe as "the maximum surveillance society" (1999) and Rule as "mass surveillance" or "the total surveillance society" (2007: 15, 162; 2009: 10) is a phenomenon typical of this modernity. It belongs to modernity not only in terms of technology but in its congruence with the rational individual's demand for the protection that surveillance provides in a hostile world which has set aside the political interpretation of the present. For this reason, surveillance research has defined one of its tasks as identifying and elucidating the conflicts of interest, and the repressive and exclusionary mechanisms, that exist under the surface of this socially neutral practice. They seek, in short, to return the question to the political sphere.

It has been posited that a central feature of video surveillance is social separation and exclusion—clearly a political issue (Hempel and Töpfer 2009; Lomell 2004; Lyon 2003; McCahill and Norris 2003; Gilliom 2001). The Foucauldian concept of *governmentality* (Borch 2005; Cameron 2004) has been the point of departure for other studies of the 'depoliticised' surveillance society. Surveillance is a model of social control in a social order characterised by governmentality rather than government (Björklund 2011). Individuals' obedience of the law is not driven primarily by traditional repressive measures or moral arguments but by establishing the conditions for choices made at the individual level. The risk of being discovered by a video camera becomes a variable that the individual must take into consideration when making calculations in the market that encompass not only traditional utility of action but social relations in general—including the decision to commit a crime.

The point of departure for this study is video surveillance as an expression of a discursive, taken-for-granted order. In one sense it is possible to talk about such an order at a universal level. Such a discourse can be found in the context of (neo-)liberal norms and ideas about the marginality of politics, ideas accepted more or less throughout the democratic world. These make the massive expansion of video surveillance, as a form of social control that belongs to an individualised social organisation, possible. Yet there are also specific national, discursive orders that interact and contrast with this over-arching surveillance framework. The universal discursive order is reflected in the national discourses, but they are, at the same time, largely unique. It is here that this study's comparison of three states serves a function. The purpose is to analyse how the discourse concerning video surveillance can assume different expressions within the practices of the (late) modern state.

Poland is a particularly interesting case; as yet, there are only few studies on Polish surveillance. We may also point out that there are relatively few comparative studies of video surveillance, especially few including former Eastern bloc countries. Nonetheless, several comparative studies concerning surveillance in general and video surveillance in particular deserve mention. One such study is Flaherty's (1989), which compares surveillance practices and legislation in five countries—Germany, Sweden, France, Canada and the United States—from the perspective of privacy. Another is a comparative study of video surveillance conducted in the years 2001–2004 by the Urbaneye Project. This study was distinguished both by its ambitious comprehensiveness and its analytic ambitions. It produced fifteen articles problematising surveillance practices of a number of European states.[1] A welcome contribution in the field is the international overview provided in the 2012 anthology *Eyes Everywhere: The Global Growth of Camera Surveillance* (Doyle et al. 2012). Finally, we must mention Gras's 2004 article, which compares national legislation on video surveillance in Europe, and Hempel and Töpfer's 2009 article on the similarity of justifications for video surveillance in England, Germany, and France. In the area of 'privacy studies', a research field of relevance to work on video surveillance, there are a number of useful, larger comparative studies. Examples include Bennett (1992), Bennett and Raab (2006) and Rule and Greenleaf (2008). This research identifies interesting national differences and similarities. What is missing, however, is systematic contextualisation. This is precisely what a discourse-analytic approach can provide. What does video surveillance express within a particular societal context?

The concept of social order used in critical discourse analysis is useful for addressing this question. According to Fairclough (2008), the social order is manifest in discourses and strengthened discursively, but the relationship can also be reversed—the manifestation of the discourse influences existing structures of control. In essence, the social order is about control, and critical discourse analysis is the branch of discourse analysis that most systematically focuses on social control and its integration in discourse (Fairclough 2008; Lazar 2005; Locke 2004; Weiss and Wodak 2003). This means that critical discourse analysis makes an analytical distinction between the discursive and non-discursive (Fairclough 2008: 54)—there is a social reality outside the discourse. Thus, "the complex interrelations between discourse and society cannot be analysed adequately unless linguistic and sociological approaches are combined" (Weiss and Wodak 2003: 7). Proponents of critical analysis argue that social reality is not entirely discursively structured. Building on these assumptions, Fairclough states that in addition to social practice and the discourse itself, analytic interest must be directed towards discursive *practices*. These are manifest in the ways in which texts are produced, communicated, and spread (Fairclough 2008: 93). The discursive practice is a link between social practice and discourse, and therefore has an impact both on the manner in which control is exercised and on the character of the discourse.

The distinction between these three discursive elements—the social order, which we will term the *social regime,* the *discourse practices*, and the *discourse* itself—is central to our analysis. Our point of departure is the mapping of the discourse practices, which are then used to reconstruct national discourses. Investigating discourse practices involves localising and characterising discourses—that is to say, describing how and where central discursive statements, documents, and debates are produced. The appearance and relative weight of discourse practices have impact on the content of the national discourse.

Discourse is a structured totality of expressions of meanings fixation in which meanings can be more or less constant (Laclau and Mouffe 2008: 157ff). We find it in the signifying concepts of key texts, documents, and debates on video surveillance, which are then placed in the context of other concepts. How do we identify the meaning-bearing concepts of a given discourse? Comparative analysis provides some help, because a concept's frequent appearance in one national context and absence in another indicates that it is a central concept in the former. In principle, however, there are two types of meaning-bearing units. The first consists of words that often appear in all discourses concerning surveillance—appearing, perhaps, more or less frequently but always present in some way in all information and debate. Examples of these are the terms *privacy, security,* and *crime prevention*. Even though these terms are used frequently, they can have quite different implications in different discourses, something that becomes clear when they are placed in a wider conceptual context. The second type of meaning-bearing term consists of discourse-specific words and expressions that repeatedly occur in a particular discourse and can therefore be seen as conveying meaning.

This study ends by sketching what may be described as an order for the national conversation on video surveillance, what we term *the social regime*, which includes characteristics of the discourse and is stamped by discourse practice. The social regime is a concept used to describe where 'power over the discourse' is located, and thus concerns the institutions and actors that direct the discourse or, at least, are central to its formation. It is a structure for conversation which implies a certain type of control over it.

Before continuing, a few things should be mentioned on the subject of locating video surveillance in a larger discursive context. Video surveillance is, of course, only one of a number of different technological tools used by states and other institutional actors. In large measure, these have overlapping implications. At the same time, different surveillance practices co-vary in such a way that it is possible to talk about a qualitatively new type of surveillance (Haggerty and Ericson 2000; Marx 2002; Hier 2003). As a general phenomenon, the technique of technological surveillance has changed social relations and the conditions for social control. Video surveillance is an example of this broader phenomenon and discourses about it should be understood as relatively open. If other types of surveillance had been included in this study, it would have influenced the appearance of the discourses, though probably not to any large extent.

THE DISCOURSE PRACTICES OF VIDEO SURVEILLANCE

How widespread is video surveillance in the three countries investigated here, and when was it established? It is difficult to say exactly how many video cameras are in operation, and this information is not always of much value, since the cameras most often are interconnected—sometimes crosswise—in different systems. Video surveillance was introduced in Germany as early as 1958 as a traffic-control measure (Hempel and Töpfer 2002: 8). Sweden followed suit several years later, installing cameras in its road network. The next step was to use film technology to protect sectors that are vulnerable to criminal activity—for example, banking and the trade in valuable objects. The great expansion of video surveillance occurred at the beginning of the 1990s, when it became a general practice in public transportation, schools, workplaces, and shopping centres. The extent to which public authorities, in particular the police, have access to surveillance material is an issue of central importance. In Sweden, with the exception of the national Road Administration (Trafikverket) cameras, there has been a restrictive attitude as regards public authorities' video surveillance in public spaces. The police do not have the authority to install permanent video cameras, except in a few locations in Stockholm's city centre. In Germany, video surveillance of public spaces by the police is strictly regulated, albeit with some variation between the various federal states. However, in both countries the police have access to material collected by the private sector. In Polish cities such as Warsaw, open-street video surveillance is carried out by the cities' own civilian surveillance bodies, but the police—a national authority—have vast amounts of unregulated access to the material (Gniadek 2009).

In general, the development in Poland differs from that of the other two countries. Video surveillance was widely introduced in Poland after the fall of communism. From then on, it came into use in most sectors of society. However, some areas, such as video surveillance in ambulances, have been subject to discussion and restriction (Pochrzest 2007). In Sweden and Germany, a more continuous legal and political process has been aimed at setting boundaries for how and which sectors can be monitored with video surveillance. These processes take place within the framework of three discursive practices: *the political-parliamentary, the judicial-bureaucratic*, and *the public*. As we shall see, a comparison of the countries shows that they differ both with respect to dominant practices and the content of the practices.

Political-parliamentary Practice

Political-parliamentary practice is exercised in arenas in which general political decisions about video surveillance are made. It is first and foremost expressed in decisions and legislative proposals announced by governments and during parliamentary debate. Thus, the first question that must be raised is to what extent political parties have an interest in politicising video surveillance. Voters' interest in the matter is quite weak. In general, people are positive to video surveillance, perceiving it as a security-promoting measure. Accordingly, for tactical, vote-seeking reasons, political parties have more incentive to accept additional surveillance than to voice critical opinions. The other question is whether views about video surveillance and surveillance in general differ according to party ideology. Surveillance questions fall largely, if not entirely clearly, along the political dimensions of 'law and order' versus 'rights', and 'progress and growth' versus 'soft, interpersonal values'. This does not mean, however, that the parties link surveillance to these dimensions of the political debate.

Political-parliamentary practice is about how lively the debate is, the grouping of different positions, and the form in which positions are communicated. A lively political practice is characterised by the packaging of opinions/understandings as party-political positions. In such cases, the question is given an ideological interpretation that distinguishes the parties from one another. Given that the basic attitude towards increased surveillance in the societies studied here is positive, the conditions for lively discourse have mainly to do with how opposition is constructed. Opposition to surveillance grows out of the wish to protect the right to privacy, in one form or another. The relevance of the political discursive practice depends on the degree to which arguments about privacy can make it into in the parliamentary arena, and what form they then take. Video surveillance has been most politicised by political parties in Germany and least in Poland; Sweden is somewhere in the middle. However, the politicisation of the issue in Germany and Sweden has occurred in somewhat different ways.

Video surveillance is not a core issue for any of the Swedish parties, nor has it been for any sitting government. Nonetheless, there has been no dearth of parliamentary debate; since the mid-1970s the issue has been repeatedly discussed in parliament. First and foremost, the debates have been a reaction to revisions and evaluations of Swedish legislation on video surveillance, which has existed for over thirty years. Technological developments and social demands have led to regular reviews of the relevant legal texts.

It is generally fair to say that since 1977, when the first legislation was passed, all political parties in the Swedish parliament have shifted to less critical views of video surveillance—something that might seem paradoxical given that the legislative changes discussed were meant to strengthen the safeguards of individual privacy. It is interesting to note that when questions of video surveillance were raised in the parliament in the 1970s and '80s, it was for the purpose of drawing attention to the risks that such activity might entail in terms of the state's exercise of power and violations of personal integrity (Regeringsproposition 1975/76:194; Regeringsproposition 1989/90:119). Lately, when the question has been raised, motions and parliamentary interpellations have almost exclusively been concerned with the desire to increase the number of video cameras in order to counteract increasing criminality.

The partisan positions in the debate have been quite unclear; they have, above all, failed to adhere to any clear party-ideological pattern. The Left (Vänsterpartiet) and Green Party (Miljöpartiet) have been most consistently critical of surveillance. During certain periods, the Social Democrats have been the most outspokenly pro-surveillance party. Here, one should mention that the question of surveillance reasonably ought to separate 'technology and progress' parties from those more concerned with protecting humanistic, 'soft' values. However, the question is not discussed in these terms. The other parliamentary parties have not been particularly active. Their contributions to the debate have been characterised by reasoned argumentation without firm position taking either for or against video surveillance. When the issue is put on the agenda, it is often more because of a local rather than a party-political interest; often, a specific municipality's member of parliament (MP) rises to explain the need for surveillance in that municipality. Accordingly, the demand for increased video surveillance is most often pursued as a matter concerning individual parliamentarians, not parties as a whole (Motion 2005/06:Ju275; Interpellation 2007/08:383; Motion 2009/10:Ju348).

Swedish parliamentary debate on video surveillance has been only weakly linked to other debates that addressed (potential) violations of privacy, and thus the broader question of and legislation on data protection. This is due to the fact that Sweden lacks a principle-driven rights-oriented approach to privacy (Björklund 2012). Rather, as we shall see, the dominant approach has highlighted a benefit specifically linked to video surveillance: crime prevention. Video surveillance has not been clearly placed in the context of data protection, nor is it explicitly discussed in the Swedish data protection law adopted by the parliament, the Personal Data Act (Personuppgiftslagen 1998:204). This has perhaps impeded the politicisation of video surveillance in Sweden, but it should also be noted that data protection is not a particularly partisan question in Sweden, either. When the first data protection law was passed in the early 1970s, there was a tendency towards a left-right politicisation of the issue (Söderlind 2009), but this has disappeared over time.

Sweden's political practice has largely formed within the framework of the tradition of official public investigations, which are complemented by expert and stakeholders' commentaries on proposed laws. These are garnered during the preliminary opinion rounds upon which proposed laws are often sent, the so-called *remiss* process. The Swedish government has initiated a number of investigations of video surveillance (Statens offentliga utredningar 2002; Statens offentliga utredningar 2009); the expert opinions produced during the remiss process have guided political debate. Thus, the dominant attitude towards video surveillance has been guided not by principles but by instrumentality.

In Germany, video surveillance is a high-order political question; it has been actively debated for several decades, and the debates have been rooted in ideological differences. Video surveillance has caused recurring disagreements in both state and federal parliaments. The salience of surveillance, and particularly video surveillance, is evident by the fact that all large major political parties publish their views on the issue on Internet homepages and blogs. The Christian Democratic Union (Christlich-Demokratische Union, CDU) has been most aggressive in promoting an expansion of video surveillance at both the state and federal levels; such policies are felt to be keeping with the party's strong stance on law and order (Töpfer et al. 2003: 22). For many years, with varying but substantial success, the CDU has promoted state legislation that will increase police authority (Töpfer et al. 2003: 22; Christlich-Demokratische Union 2003; 2009). The CDU faction is also strongly engaged in the Berlin state parliament in promoting video surveillance in public transport, such as the subway system (Abgeordnetenhaus Berlin 2007). In general, the CDU can be counted as a consistent supporter of video surveillance, whether carried out by police, other public authority, or private actors. This accords with the party's ideology; the CDU has, moreover, expended a good deal of effort in strategic alliances with industrial interests and nonparty authorities felt to be sympathetic to its political agenda (Töpfer et al. 2003: 22; Svenonius 2011; see also Töpfer, this volume).

The position of the German Social Democratic Party (Sozialdemokratische Partei Deutschlands, SPD), on the other hand, has been less consistent. The party has participated in efforts to facilitate video surveillance, but with varying degrees of reluctance. In the early years of the new century, the CDU and SPD joined in recommending police surveillance of high-crime areas. This decision opened the door to revisions of the federal states' police legislation, allowing the police increased authority to use video surveillance in their work (Töpfer et al. 2003: 22). In the federal state of Berlin, however, the SPD endorsed a restrictive policy; here they opposed the CDU's repeated proposals to revise police legislation. In 2007, when the SPD led the Berlin government, the party switched positions; amended legislation was passed

(Abgeordnetenhaus Berlin 2007). This measure was supported by the Left (Die Linke), which shared government with the SPD, despite its essential incongruence with the Left's position on the question in other contexts. The Greens (Die Grünen) were and still remain the most consistent critics of expanded video surveillance.

Accordingly, although video surveillance is an ideological question in Germany, it is not an unequivocally partisan issue. Rather, it is ideological in that it is debated against a background of ideological principles. Video surveillance is integrated into a larger debate about data protection which, in Germany, is clearly connected to democracy and citizens' rights. As in Sweden, the debate is largely concerned with the benefits of video surveillance. However (as we shall see), in contrast to their Swedish counterparts, German actors also formulate the issue in terms of concepts of society and society's future. The protection of personal information—that is, the right to control information collected about oneself—is here a central question.

The German debate on data protection has been intense; the issue has a complicated, politicised history. The first national law on the issue, the Federal Data Protection Act (Bundesdatenschutzgesetz), was passed in 1977 after a long, ideologically coloured debate concerning regional political interests. From the outset, data protection was formulated in terms of citizens' rights, an issue on which parties placed differing emphasis. The question was also formulated as a disagreement between the interests of representative democracy and the interests of the bureaucracy. The Left argued in favour of legislation that would set up a Data Protection Authority anchored, institutionally, in the federal parliament (Bundestag). The German bureaucracy, with its traditions of integrity and legalism, opposed this. The CDU was also strongly opposed to parliamentary influence over data protection, arguing that this would constitute an anomaly in the German system (Bennett 1992: 211ff). The current legislation, however, clearly connects the bureaucracy of data protection with the political arena. Every other year, the German Commissioner for Data Protection and the Freedom of Information (Bundesbeauftragte für den Datenschutz und Informationsfreiheit) is required to submit specially formulated activity reports to the federal parliament. The parliament can also request the regulatory agency to provide expert opinions and reports on various matters (Bundesdatenschutzgesetz, Section 26; see Kilian 2008: 83;). These sessions guarantee that data protection remains current on the political agenda; political practice is linked to judicial-bureaucratic practice.

Germany's political-parliamentary practice concerning video surveillance is related to a broader debate about data protection. It is also connected to a constitutional guarantee of the right to a private sphere. A declaration of principle concerning video surveillance can be found in the federal German Data Protection Law. This declaration is, in turn, based on the constitution and interpretations thereof. Data protection and video surveillance have both been deeply rooted in a discourse on citizens' rights.

Within this framework, a clear political opposition has been formulated and manifested—not least in the parliamentary arenas.

In Poland, video surveillance has not been articulated as a party-political question at all. There is no national legislation on the subject, nor has the government brought up the issue for debate in the parliament. So far, at least, video surveillance has remained apolitical. It should, however, be noted that the Polish legislation on data protection, and Poland's constitutional provisions, in fact provide the potential for a political discourse in the form of clearly expressed civil rights. These could be used to make video surveillance an ideological matter. (We will return to this below, in our discussion of the Polish discourse.) However, Polish data protection legislation does not explicitly mention video surveillance. To be sure, the Polish data inspection authority is required to submit regular reports to parliament, but these reports are not given the same attention as they are in Germany.

The Swedes and Germans embed video surveillance in political-parliamentary practices. This is not the case in Poland, there the content of the discourse is not formulated by a political authority, either in government or in the parliamentary arena. The discussion of video surveillance is driven by a logic that has not been formulated by those with political power. Of the three countries covered by this research, Germany's video surveillance discourse is most clearly formed by political practice. There, video surveillance is part of an ideological discourse that divides parties (more or less consistently) according to their views on the relative weight to be given law-and-order issues as opposed to civil rights. This is not true of Sweden, even though video surveillance is regularly discussed in Swedish parliamentary forums. There, party-political tension is largely linked to questions concerning the relative instrumental benefits of video surveillance. The lack of ideological discussion and political debate leaves considerable space for the exercise of judicial-bureaucratic practice; in the absence of party-political disagreements at the government level, the question is defined as an administrative matter. In Germany, when it comes to data protection—including video surveillance—the political and the judicial-bureaucratic spheres are communicating vessels, but data protection has also been the object of a power struggle between politics and bureaucracy. Whether one or the other owns discursive practice is an open question. There is also a significant German public practice that supports the political, something which is glaringly absent in Sweden and Poland. We will return to this point below.

Judicial-bureaucratic Practice

The core of judicial-bureaucratic practice is, on the one hand, the exercise of authority and judicial review and, on the other, the efficient handling of cases. This practice is, unsurprisingly, very important to video surveillance.

Such surveillance involves a policy issue which, if regulated, must lead to bureaucratic and legislative case-handling activity. However, the significance of the judicial-bureaucratic sphere as discursive practice varies according to the scope of other practices. An absence of political practice concerning video surveillance can, as noted above, increase the importance of judicial-bureaucratic practice. Another thing that varies according to country is the relative importance of the judicial sphere vis-à-vis other parts of public administration. Is video surveillance, one may ask, largely a judicial question or primarily a bureaucratic concern?

By and large, it is fair to say that the judicial-bureaucratic practices of our three countries differ significantly. The national data inspection authorities are important actors in all three, though to different degrees and in different ways. But there are, further, numerous nationally specific institutional solutions for dealing with the policy area constituted by video surveillance.

In all three states, as in the entire European Union, data inspection authorities are endowed with largely similar authority and responsibilities. At the end of the 1990s, the EU's Data Protection Directive 95/46/EC obligated all member states' governments to adopt national laws to prevent the inappropriate handling and use of personal data information. Some parts of the directive are mandatory: each member country is, for instance, obliged to establish specific institutions to ensure respect for data protection. However, there is space for national variation. Country-specific factors such as legal traditions and public opinion affected the formulation of data protection in different European nations. An important factor was, of course, the existence of a national structure for data protection before the passage of the EU directive, as was the case in Sweden and Germany.

When we compare the Polish, German, and Swedish data inspection authorities, we can see that, in formal terms, the Polish inspection authority is the most independent vis-à-vis the government. The parliament appoints the Polish Inspector-General for Data Protection. The federal German data inspection authority, the Authority for Data Protection and Informational Self-determination, is, by contrast, closely tied to the government, unusually so when compared to other administrative units in German public administration. The government nominates a candidate for the position of federal commissioner, whose appointments are confirmed by parliamentary vote. Finally, he or she is appointed by the president. What is most important, however, is the fact that the German data protection authority is placed under the Ministry of the Interior and monitored by the government. This system of organisation ensures that surveillance retains political relevance. The organisation of the Swedish Data Inspection Board (Datainspektionen), finally, conforms to the traditional Swedish public administration structure, whose organisation can, in practice, be decided anew by each government. In this, the Swedish Board is not qualitatively different from other parts of the public bureaucracy.

The first Swedish national data protection law (Datalagen 1973:289) dates from 1973, and created the Data Inspection Board. The amendments required by the EU directive led to the 1998 Personal Data Act (Personuppgiftslagen 1998:204), which came into force in 2000. In practice, the new law decreased the power of the Data Inspection Board. Under the first set of legislation, the board was responsible for approving the registration of personal data; this, indeed, was its main function. No computer-based registration could take place without the board's approval. Today, the Swedish Data Inspection Board issues no such licenses. Rather, its most important responsibility is to monitor compliance with the law and to provide information and training (Datainspektionen 2011a). The distinction between 'structured' and 'unstructured' surveillance material, added to the law in 2007, further limited the power of the Data Inspection Board. In principle, the Personal Data Act is no longer relevant to so-called unstructured data registration, as long as such registration cannot be shown to violate anyone's personal integrity. An example of such unstructured registering is video surveillance, in cases when the material is neither processed nor saved for a significant period of time. This means that the Personal Data Act has limited relevance as far as video surveillance is concerned.

In Sweden, the Data Inspection Board does not make the most important regulatory decisions concerning video surveillance. This power resides in the Swedish county administrative boards (*länsstyrelser*). Since the end of the 1970s, however, the county administration boards are responsible for granting approval in cases of video surveillance. In contrast to Poland and Germany, Sweden passed a specific video surveillance law as early as 1977; it is presently known as the Act on Public Video Surveillance (Lagen om allmän kameraövervakning 1998:150).[2] The core of the law is the so-called proportionality principle, which mandates that the benefits of video surveillance be weighed against the violation of individual privacy. Although this video surveillance law takes precedence over the Personal Data Act, it only covers surveillance of public places or places to which the public has access. This gives the data protection law residual relevance when it comes to 'structured' surveillance of other spaces, such as apartment buildings and workplaces.

The county administrative boards' mandate has been written into the Act on Public Video Surveillance. Sweden's twenty-one county boards are responsible for ensuring that the law is followed and for issuing licenses for video surveillance. With some exceptions,[3] anyone who wants to conduct video surveillance must apply to the county board for permission. This might seem to furnish evidence both of Sweden's devotion to privacy and its efficient safeguards against unmotivated video surveillance, especially when coupled to the national law that specifically regulates video surveillance. But this is not the case. In practice, Sweden's county administrative boards have neither the mandate nor the competence to act pro-actively in questions of individual privacy. They are irrelevant to national-level policy formation—that is, the level on which the national data protection authorities act. The county boards are charged only with implementing national policy. Individual county boards have the freedom, in their capacity as license issuers, to interpret the law more or less restrictively. In practice, however, heavy workloads often lead county administration boards to approve applications as a matter of routine. Applications are renewed without a new evaluation of the arguments for surveillance, even though circumstances may have changed (see Loftsson, this volume).

In order to maintain regional conformity, the Attorney General (Justitiekanslern) is authorised to appeal the decisions of the county boards to the administrative court of appeals (*förvaltningsrätten*).[4] The Attorney General has taken up a number of cases, including surveillance in taxis, churches, hospitals, parking garages, and apartment buildings Some cases have been appealed up to the highest judicial level, the Supreme Administrative Court (Regeringsrätten 6572–98 09/29/2000; Regeringsrätten 5767–02 03/09/2004; Regeringsrätten 1450–01 09/26/2001; Regeringsrätten 6462–05 02/02/2009; Regeringsrätten 7873–08 02/11/2010; Högsta förvaltningsdomstolen 4459–10 01/31/2011). A county administrative board decision to allow the Stockholm Police Authority (*polisen*) to install video cameras has also been appealed (Regeringsrätten 7873–08 02/11/2010). In general, these judicial processes have been defined by the so-called proportionality principle—that is, the requirement that the need for or interest in surveillance be weighed against possible violations of privacy.

This is the sort of assessment the county administrative boards are required to make under Section 6 of the Act on Public Video Surveillance. Accordingly, the legal process does not introduce a qualitatively different perspective on whether or not video surveillance should be permitted. The county board can only decide whether a given factor or factors should carry greater or lesser weight. If the purpose is to prevent crime and accidents, then the board is supposed to be less restrictive. This is in keeping with the intention of the law, outlined in the preparatory documents composed while the law was being drafted (Regeringsrätten 7873–08 02/11/2010).

As mentioned above, as far as video surveillance is concerned, the role played by the Swedish Data Inspection Board is very limited. The board does, however, have some relevance when it comes to video surveillance in areas not open to the public, such as schools. These locations are not covered by the video surveillance law, and, between 2005 and 2010, the Data Inspection Board surveyed cases concerning school surveillance. The ensuing judicial processes treated video surveillance as a matter solely of proportionality (that is, in terms of the strength of a legitimate need as weighted against the individual's privacy interests—in this case, those of pupils and employees; Datainspektionen 2008). Thus, like other Swedish actors, the Data Inspection Board raises no additional principal arguments in favour of protecting personal integrity. An information folder published by the board makes it clear, indeed, that the central task of Swedish data protection is to weigh interests (Datainspektionen 2009).

The German data inspection authority has a far stronger position than does the Swedish authority. The world's first administrative structure for data protection was established in Hessen in the early 1970s. This German state had adopted its first data protection law in 1970 (Rule and Kilian 2008: 83). The Federal Data Protection Act was not passed until 1977. The work of the Commissioner for Data Protection and the Freedom of Information is, in fact, circumscribed by the fact that all German federal states have their own data protection laws. The federal authority is required to cooperate with the data protection authorities at state level. However, the German data protection authority has a much more ambitiously conceived field of operation than do its Polish and Swedish counterparts. Data protection permeates both German public administration and private industry by dint of the special data protection ombudsmen that the law requires of all organisations above a certain size. The German data protection authority functions as an advisory and corrective body towards these ombudsmen—and towards public bodies in general. Chapter 3; section 25 in the Data Protection Law states that the federal commissioner shall—if he or she finds that the act is violated—lodge a complaint with the competent supreme federal authority. In contrast, again, to the Swedish case, the federal-level law on national data protection includes a clear mandate regarding video surveillance of public places (or, to paraphrase the legal phrase, places to which the public has access). Video surveillance is expressly mentioned in the Federal Data Protection Act.

Proportionality is also an important principle in the German practice, but the country's data protection authorities do not regard its fulfilment as their primary responsibility. Their first task is to enforce efficient data protection and to guard privacy (BfDI 2007c; 2010). Citizens' rights are central. The origin of German data protection is the constitutionally motivated principle of informational self-determination (BfDI 2007b; 2011a). The German authority is also called upon to be active in legislative matters and to articulate arguments and recommendations in political forums, not least through its regularly issued reports to state and federal parliaments. Not infrequently, this leads to differences of opinion between the data protection authorities and the political assemblies (Abgeordnetenhaus Berlin 2010).

A number of cases that deal with surveillance and video surveillance have been decided by the German judicial system. Verdicts have been issued on, among other things, video surveillance of workplaces, public places, and traffic. Some have also been heard by the Constitutional Court, including a 2007 case concerning video surveillance of medieval synagogue ruins in a German city (Bundesverfassungsgericht 1 BvR 2368/06 02/23/2007) and a 2010 case on video surveillance of traffic violations (Bundesverfassungsgericht 2 BvR 1447/10 08/12/2010). The Constitutional Court is meant to correct legislation and measures that threaten the protection of the private sphere.

In Germany, video surveillance is regulated by data protection legislation, but also, and more exhaustively, in supplementary legislation and in the regulation of various organisations. In addition, a large number of legal instruments meant to regulate other types of surveillance have been adopted over the last decade, several of which have relevance to video surveillance (see Töpfer, this volume). The data protection law is complemented, very importantly, by the quite restrictive legal stipulations of state police laws. As the German discourse so clearly springs from the idea of a state governed by law, it is natural that police video surveillance be subjected to special legislation. First and foremost, state institutions must be well supervised. Here, Germany and Sweden differ; in the latter, police video surveillance is regulated by the general, national video surveillance law (except in cases of secret video surveillance).

The Act on Personal Data Protection (Ustawa o Ochrony Danych Osobowych) is Poland's most important regulatory document for technological surveillance. The law was adopted in 1997, a few years before Poland became a member of the European Union. In 2010, the activities of the Polish Inspector General for Personal Data Protection (Generalny Inspektor Ochrony Danych Osobowych, GIODO) were still in a phase of consolidation. At that point, video surveillance was a new, but not particularly central, question on the authority's agenda (Krasinska 2009). This Polish data inspection authority serves a regulatory function regarding video surveillance; it is empowered to take disciplinary action. If video surveillance entails the processing of data—that is, if data is organised in a structured and searchable database—the data inspection authority must be informed.

GIODO also operates under a clearly constitutional mandate (Central and Eastern Europe Data Protection Authorities 2011). The right to personal integrity is inscribed in the Polish constitution and Civil Code. Accordingly, article 1:1 of the Act on Personal Data Protection opens with the words, "All individuals have the right to protection of personal information". No such formulation exists in either the Swedish or German counterparts. However, it remains untested whether or not this declaration in the Polish law has had any practical impact on video surveillance or other activities posing threats to personal integrity.

There is no specific national regulation of video surveillance in Poland, which means that the supervisory structure is a bit confused. Decisions about video surveillance are usually made locally, or are tied to various sectors and carried through by the operation in question. Video surveillance is regulated by the sector-specific laws and internal instructions of different public institutions. According to Polish Police Act, the police may use video surveillance material collected by other public institutions for the purpose of conducting preliminary investigations, although they are supposed to heed the data protection law when doing so (Ustawa o Policiji 1:14:4–5). At the national level, regulations and instructions also give the ministries of defence, finance, and interior some degree of influence over the practice of video surveillance (Waglowski 2008a).

In Poland, a large proportion of public surveillance, including that requested by police, is carried out by municipal administrations' own civil security-guard organisations, the so-called *Straż Miejska*. In Warsaw, for example, video surveillance is regulated by this security-guard institution's internal instructions, but other levels of the city administration also have influence on surveillance practice. An administrative unit for video surveillance (Zakładu Obsługi Systemu Monitoringu) is under the city's jurisdiction, subordinate to the security department—the Bureau for Security and Crisis Management (Biura Bezpieczeństwa i Zarządzania Kryzysowego). However, surveillance questions are also handled by the local police organisation. Thus, for instance, the Warsaw police force has its own surveillance centre (Centrum Monitoringu w Komendzie Stołecznej Policji). In Poland, thus, actors at different levels have different kinds of surveillance authority. How these powers are related to each other is not entirely clear. The lack of coordination makes it difficult to determine where exactly authoritative decisions are to be made; critics complain that this leaves bureaucratic actors too much operational leeway (Waglowski 2008b).

We can sum up by characterising Swedish discourse practice as endorsing a bureaucratic approach. This is expressed in the county boards' licensing activity, but also in the strong focus on proportionality that is both characteristic of such licensing and decisive in Swedish judicial processes. The proportionality principle in the Swedish video surveillance law prescribes that surveillance be handled administratively, as conflicts of interest to be reviewed at different levels of the judicial system. This stands in sharp contrast to German legalism (Flaherty 1989: 86). In Germany, video surveillance is a judicial question of rights; it extends to constitutional issues. Video surveillance is held to have an impact on citizens' rights, and thus tie into issues of democracy and the rule of law. The rights perspective gives video surveillance its particular constitutional relevance. This helps explain why video surveillance in Germany is more clearly ideologised than it is in Sweden. At the time of writing, Poland is still without a legal practice concerning video surveillance; nor does it have effective legislation for its regulation. The data protection law has some impact, but it is weak as judicial practice, and it is as yet unclear to what extent GIODO is empowered to deal with the issue of video surveillance. Local practices have developed, creating a space for different types of actors and discourses—none of which are conceived as being politically ideological. The Swedish Data Inspection Board has likewise little impact in matters of video surveillance; indeed, its mandate is generally weak. Its contributions are negligible, with the one exception of the bureaucratic discourse practice found by the county administrative boards. Operating from the perspective of citizens' rights, the German Commissioner for Data Protection and the Freedom of Information is, by contrast, an important actor in the German discourse, even if this does not always affect the actual development of surveillance. It should be noted that German legalism has, in practice, not led to more restrictions on video surveillance.

The Public Practice

The public practice is that which operates in forums outside the formal political arena (e.g., the parliament). In such forums, the three countries each display discourses more or less critical of the prevailing degree of video surveillance. These discourses are often sponsored by organisations and persons active in civic society. They spread their views through a variety of channels, including popular mobilisation, mass media, and the Internet. Public practice can also be expressed more generally in public opinion as assessed, for instance, in different types of public surveys. In Germany and Sweden, citizens' attitudes towards video surveillance have been investigated through public opinion surveys posing specific questions on the issue. The responses have shown a primarily positive attitude (Bjereld and Demker 2009). There is no reason to believe that public opinion differs significantly in Poland, though no studies have been done there.

Sweden lacks any significant civic-society activity focused on video surveillance. A few years ago, with the founding of the so-called Pirate Party (a civil rights party whose name refers to pirate downloading), the issue might have been placed on the public agenda; but this did not happen. The Pirate Party, now with a seat in the European Parliament, has focused on other questions. Germany, by contrast, hosts organisations that prioritise questions pertaining to the protection of civil rights and who therefore monitor the development of video surveillance. Some of these are purely activist in nature; others are rooted in more intellectual circles. The latter include the well-established Humanist Union (Humanistische Union), which has engaged in debates on video surveillance, and, more recently, the very controversial EU Data Retention Directive (Bennett 2008: 40). One should also mention here the researchers who treat surveillance from a critical discourse perspective and who appeal to a wider public. These include the Research Network on Monitoring Technology, and Control (Forschungsnetzwerk zu Überwachung, Technologie und Kontrolle), a group that publishes criticism and critical commentary on the theme of video surveillance on the Internet. German organisations have also used traditional methods, such as demonstrations and civil disobedience, in attempts to influence public opinion. In the autumn of 2007, for instance, a demonstration protesting surveillance marched in support of "Freedom, Not Fear" (Freiheit statt Angst), attracting participants from more than thirty German cities as well as from foreign countries (European Digital Rights 2008). Several German citizens' rights activist groups take a stance on video surveillance, as well (Hempel and Töpfer 2004: 24). As early as 2000, around 2,500 people demonstrated in Leipzig against a pilot project for video surveillance, a demonstration that ended in violence (Hempel and Töpfer 2002: 13).

There is organised opposition in Poland as well, and this, also, is nourished by a civil rights discourse. The Polish opposition occurs, however, on a much smaller scale than in Germany, and is not activist in the traditional sense. It manifests itself primarily through Internet postings, written by a few loosely organised people particularly knowledgeable in the area (Panoptykon Fundacja 2011; Waglowski 2011). It may be noted, incidentally, that the Polish data protection authority does not seem to be held in especially high esteem by those who write this public criticism (Panoptykon 2009).

Traditional media coverage of video surveillance—for example, in daily newspapers—varies from country to country. German media coverage of video surveillance is extensive. Articles take a variety of perspectives, ranging from the sensationalist to critical, supportive, or purely factual coverage. Video surveillance is, clearly, of much greater media concern in Germany than in the other two countries. In Sweden, the mass media is not a leading forum for arguments concerning the pros and cons of video surveillance. However, over the past ten years there has been a certain amount of reporting on, for example, court decisions and the expansion of video surveillance in general. Individual columnists have occasionally raised the question of whether video surveillance is compatible with the principles of a democratic society. However, this sort of article is rare. The Swedish Data Inspection Board, as well as state bureaucrats and judicial experts, have occasionally made use of the mass media to spread their views (Gräslund 2005; 2008; Strömberg et al. 2006). Carlsson (2009: 149) discusses the roll of experts and political elites in the Swedish discourse on surveillance (including video surveillance). He notes that the media gives these elites a good deal of exposure, which might impress the public with the apparently unavoidable need of expanded surveillance.

Poland's mass media reporting on surveillance is different; the issue, when broached, is raised in articles with sensationalist overtones. In the few articles that do appear, the video surveillance issue is framed by themes such as criminality and the abuse of bureaucratic power. Thus, for instance, there are articles denouncing the incompetence of the security-guard organisation that operates the cameras for misusing equipment so that pictures are leaked to the media; other articles have uncovered scandalous bonus programs for reporting crimes being filmed by video cameras. In short, Polish mass media coverage problematises specific practices rather than the activity in itself (Pochrzest 2007; Machajski 2008). However, some articles call attention to the risk that video surveillance, through its control of information, might violate the spirit of democracy (Krajewski 2007).

The public arena is shaped by the character of the policy which is being debated. That means that surveillance-critical organisations must position themselves in relation to a more or less surveillance-positive establishment. They must respond to particular arguments, which, in turn, influence the discussion's framework, including the actors and arenas to which critics must relate. Sweden is a country in which video surveillance has met little or no public criticism despite the rapid expansion in its use. This is because the Swedish discourse provides little purchase for public criticism; the controversial elements are not clear-cut. To engage in organised criticism of a bureaucratic practice without having the support of a relevant political-ideological framework is difficult.[5] The framework of German discourse gives more space to active opposition, formulated as a fight for citizens' rights. There is, moreover, an intellectual establishment that has taken a standpoint that can be reused in public criticism; intellectuals and academics also figure prominently in public debate. In this regard, Poland resembles Germany more than Sweden, although the criticism formulated in Poland's civic society is very tentative. In addition, despite its point of departure in a rights discourse, the Polish criticism lacks a clear opponent against which to direct its arguments. The positive attitude towards video surveillance permeates society without this being clearly enunciated.

THREE DISCOURSES ABOUT SURVEILLANCE—BALANCING INTERESTS, MODERNISATION, AND CITIZENS' RIGHTS

Sweden and the Discourse of Balancing Interests

The national discourses concerning video surveillance are organized around the terms that govern the various statements made in the debate. The Swedish terminology includes *crime prevention* and *balance*, as well as terms associated with some sort of benefit, such as *interest* and *need*. The central terminology also includes *integritet* in the sense of something that should be protected from violation; but it is not an association with frequent cautions against video surveillance's threat to integrity that places *integritet* at the centre of the discussion. It is, rather, the vague and shifting meaning of the Swedish term *integritet* that forms the discourse, and distinguishes the Swedish discourse from its German and Polish counterparts. The Swedish word *integritet* is usually translated into English as 'privacy' (within the surveillance discourse). However, as will be discussed below, although the meaning of the terms *integritet* and *privacy* overlap, they do not entirely coincide.

All discourses about surveillance contain a conflict between two values, expressed by each nation's respective terms for *privacy* and *security* (often also defined differently, in each context). It is the potential conflict between these that gives the question of surveillance its characteristic edge. Both concepts have positive connotations, yet in most cases they are perceived as incompatible. The way in which the contradiction is defined may differ according to national contexts and according to the specific terms' meanings. In the Swedish context, national discourse has prioritised the term *crime prevention* over the concept of security. Security as an argument for surveillance appears only in more narrow contexts—for example, surveillance in public transportation (Svenonius 2011). *Integritet* is, as we will see, a concept that discursively reduces the conflicts between the values mentioned above.

During the first decade of the new millennium, Swedish arguments in favour of video surveillance have strongly emphasised its utility in preventing crime. In government proposals (Justitiedepartmentet 2001:53; 2008:22; Riksdagsprotokoll 2006/07:1463), parliamentary debates (Motion 2005/06:Ju273; Riksdagsprotokoll 2006/07:132; Motion 2007/08: Ju417; Motion 2009/10:Ju405) and official public reports (Statens offentliga utredningar 2002; Statens offentliga utredningar 2009), crime prevention is frequently cited in urging expanded video surveillance. Other arguments have been advanced, including surveillance as a source of material evidence that will simplify judicial procedures and as an aid in preventing accidents, but crime prevention has dominated.

The effect of video surveillance in preventing crime is written into Section 6 of Sweden's Act on Public Video Surveillance, which reads, "In evaluating the use of public video surveillance, the need for surveillance to avert crime, prevent accidents and achieve other comparable goals should be given special attention."[6] This means that even those judicial authorities charged with ruling in cases concerning video surveillance must take a stance on its relation to crime prevention. In a court case of 2008–10 involving the police's use of stationary video cameras in a number of places in Stockholm's inner city, the Supreme Administrative Court ruled in favour of the police. In the decision, the court expressly referred to the police's argument that surveillance was meant to prevent crime, pointing out that this should be given special weight in the court's decision (Regeringsrätten 7873–08 02/11/2010). Crime prevention becomes part of the court's reasoning; as in the law text cited above, indeed, it simply takes for granted that video surveillance is an effective means of preventing crime.

Although the hope that such surveillance will systematically prevent crime has dominated the recent Swedish debate on the issue, the crime prevention argument is in fact relatively new in the history of Swedish video surveillance. When the first cameras were installed in the 1970s, their connection to crime prevention was weak. Generally, there was great scepticism towards the idea of video surveillance. Earlier legislation in this area, such as the 1977 Act on TV Surveillance (1977:20) and the 1990 Act on Surveillance Cameras (1990:484), was motivated by a growing concern about the threat that technical development might pose to people's privacy (Regeringsproposition 1975/76:194; 1989/90:119; Statens offentliga utredningar 1974). In its most recent form as the Act on Public Video Surveillance (1998:150), the Swedish legislation had been modified in the interests of crime prevention. The main argument behind the 1998 revision of the law was that video surveillance could be an instrument for social good. In preparatory work and parliamentary debate on the law, considerable time was devoted to discussing the advantages of video surveillance and how it could be used to prevent crime (Regeringsproposition 1997/98:64; Riksdagsprotokoll 1997/98:86; Motion 1997/98:Ju21; Motion 1997/98:Ju910).

The idea that crime prevention is a social and economic measure, meant to help those socially vulnerable groups who risk falling into criminality, has long been an important component of Swedish crime policy. It was part of the politics of social democratic welfare (Tham 2002). However, such 'socially directed' crime prevention is different from crime prevention which is 'situation-directed' (Zedner 2007: 262); it is the latter that is linked to video surveillance. As the term implies, situation-directed policies are meant to prevent criminal activity not by a particular group but in a particular situation. The situation in question is often a place where crime frequently occurs, such as a parking lot or a public footpath. Situational crime prevention rests on the assumption that the perpetrator is a normal, 'honourable' person who succumbs to a momentary temptation—the opportunity making the thief (Garland 2000: 2, Fussey 2008: 123). Situational crime prevention is radically different from social prevention. It is based on a different view of both crime and the criminal. In principle, crime becomes a normal, rational action, like any other.

It is fair to say that video surveillance is a technique that is almost perfectly suited to situational prevention (Waples et al. 2009). It cannot be used for social crime prevention. It is also apparent that the normative premises upon which situational crime prevention rests have become part of the Swedish discourse. In the 2002 study conducted to evaluate Swedish video surveillance, criminality was, in fact, defined as a result of rational choice, rather similar to people's other everyday decisions: "The point of departure for situational crime prevention is that crime is rational behaviour that is preceded by a decision to commit, or not commit, a crime in a certain situation" (Statens offentliga utredningar 2002: 98). We can all fall for the temptation to commit a crime if the opportunity arises, if we stand to benefit from the act. The role of video surveillance is to influence our calculations. The increased risk of being discovered is, it is thought, a factor that can tip the balance in favour of obeying the law.

Swedish law requires the posting of signs, clearly visible, that inform people that they are being monitored by video (Lagen om allmän kameraövervakning 1998:150 §3). This requirement has an important implication for situation-directed crime prevention. After all, rational decision making requires full information about the risks and benefits associated with the alternatives. Putting up signs can also reduce the risk of violations of personal privacy. The logic here is that those who consider surveillance an invasion of privacy can avoid video cameras if signs tell them where they are placed. The degree to which privacy is violated may also be deemed less if people are informed that they are being filmed. Further, if a sign advertises the presence of a camera, and people chose to pass by it, this can be seen as consent; there is no violation. This view is reflected in the fact that Swedish legislation is very restrictive when it comes to secret video surveillance. However, a strong justification for the Swedish requirement that signs announce the presence of video cameras is situational crime prevention. If the public is unaware of video surveillance, the technique is useless as a deterrent (particularly since most Swedish surveillance cameras are unmanned). It is the camera itself—or, rather, knowledge that the camera is there—that must do the job; there is very seldom anyone behind the camera to react. In Norris and Armstrong's term, video surveillance is an "automatic functioning of power" (1999: 92).

Situational crime prevention is also closely tied to the idea of 'public' video surveillance. The concept of situational prevention was introduced at the same time that the term '*public*' (det allmänna) was added to the legislation. In 1998, previous legislation in this area—the Act on Surveillance Cameras (which replaced the Act on Television Surveillance)—was redefined as a law on 'public' video surveillance. *Det allmänna* is a Swedish concept that is difficult to translate into other languages. It refers to a domain between the state and the individual, and is close to—but not the same as—'the social'; it includes the notion of mutual collective responsibility (Jacobsson 2010). *Allmän* covers both public space (det allmänna utrymmet) and the public (allmänheten).

The Swedish parliamentary debate at the end of the first decade of the new millennium illustrates how the connection between video surveillance, situational prevention, and surveillance of the public—that is, the common person—is rooted in the discourse. This was most clearly enunciated by a Social Democratic MP: "People hesitate to snatch a purse if they know that there are cameras" (Riksdagsprotokoll 2006/07:132). Representatives of other parties also use expressions that show a broad acceptance of the idea of video surveillance as a general deterrent—that is, directed toward the general public—against criminal action (Riksdagsprotokoll 2007/08:69; 2006/07:132; Motion 2005/06:Ju273).

Situational crime prevention puts great faith in video technology. This is naturally a problem if the technology does not live up to expectations. There is, today, an extensive debate as to whether video cameras actually prevent crime. It has been established, both in international and Swedish research, that video surveillance does not prevent all types of crime (Brottsförebyggande rådet 2003; 2007). There is, in Sweden, increasing awareness of some notable weaknesses in the empirical foundations of arguments underlying many comprehensive governmental-level declarations and documents endorsing expanded video surveillance. It has not always been possible to unequivocally confirm its effectiveness. Therefore, in answer, the argument has sometimes been turned around: the effectivity of video surveillance has been taken for granted, and the question posed, rather, as having only to do with defining the areas in which it can be most appropriately used. In the Ministry of Justice's instructions concerning preparations for a new video surveillance law in 2001, the preparatory committee was urged, simply, to make proposals about how video surveillance could be used in a strategic and effective way to fight crime (Justitiedepartementet 2001:53).

Video surveillance as an instrument of crime prevention is a central theme in Swedish discourse. However, such instruments must be weighed against the good of privacy—in Swedish *integritet* (integrity). The exact meaning of the term *integritet* is seldom explained. There are few efforts to define the term, or to describe the ways in which privacy might become endangered. This vagueness is in part caused by language and/or cultural factors. As noted above, Swedish lacks a thoroughly useful word with which to describe that which surveillance might threaten. The English word *privacy* has no exact counterpart in Swedish. The Swedish word *integritet* translates into both the English word *privacy* and the English word *integrity*, but is nonetheless used as a synonym for privacy. In order to increase clarity, the phrase *personal integrity* is sometimes used, but this is hardly a meaningful addition because integrity cannot be anything but personal. At the same time, because of this terminological ambiguity, the meaning of the concept is shifting.

There is no strong tradition of defining, or protecting, personal privacy in the Swedish legal history. Neither the protection of data nor of privacy is mentioned as pertaining to an individual as a right. Therefore, privacy has no special status in relation to other demands the citizen can make on the state. Nor is the Swedish law on data protection based on an unambiguous rights perspective.[7] This is consonant with the fact that Swedish constitutional documents provide a relatively weak protection of privacy. In the Swedish case, privacy is not expressly mentioned as a public, inviolable right (see Jonason, this volume). The 2010 revision of the constitution's second chapter, Section 6, states that each and every person is, "vis-à-vis the public, protected from significant invasions of personal integrity, if this occurs without consent and implies surveillance or mapping of the individual's personal situation". This formulation is stronger than the text it replaced, but remains imprecise and limited when compared to the protection given privacy in other countries.

When personal privacy is discussed in connection with video surveillance, it is usually defined as an interest—that is, not a right—that may be in conflict with the interest of fighting crime. The formulation in Section 6 of the Act on Public Video Surveillance states, "Permission for public video surveillance is to be given if the interest of the surveillance weighs heavier than the individual's interest in not being the object of surveillance" (Lagen om allmän kameraövervakning 1998:150 §6).

That personal privacy is formulated as "the individual's interest in not being the object of surveillance" is significant for the Swedish discourse, in both its political and judicial-bureaucratic practices (Regeringsrätten 7873–08 02/11/2010; Datainspektionen 2011b; Datainspektionen 2008; Lambertz 2008; Riksdagsprotokoll 2006/07:1463; Justitieutskottets betänkande 1997/98:JuU14; Riksdagsprotokoll 1997/98:86; Motion 1997/98:Ju20; Motion 1997/98:Ju22; Motion 2007/08: Ju417). The term *interest* belongs to a utilitarian terminology, one aimed at balancing different kinds of utilities. A utilitarian perspective makes it easier to relativise values, to weigh one against another. The view that it is possible to effectively balance different values demands, or is at least facilitated by, a utility-maximizing perspective. The value of an interest can, moreover, be heightened or diminished. If personal privacy is understood as a civil or citizen right, as is common in other nations, it is much more difficult to weigh it against other goods. Rights are absolute in a way that utilities are not, and thus not so easily amenable to balancing.

The notion of a balance between different interests is one of the foundations of the Swedish discourse on video surveillance. Legislation stipulates the idea of a balance; the balance of interests is a well-established point of departure for both proponents and critics of expanded surveillance. Naturally, the notion of a balance between values is also used within the international discourse on security (Raab 1999: 69; Liberatore 2007: 114), which likewise must weigh possible conflict between security and the right to privacy. This, too, is true of all three countries being studied here, but only in Sweden is the concept of *balanced interests* a fundamental pillar of the discourse. This accords well with a legal tradition that has de-emphasized constitutional guarantees of individual rights.

The notion of balance—'the proportionality principle', to use the judicial phrase—appears both in legal and political practices. Its roots lie in terms used to describe how problems are resolved by management—that is, in a bureaucratic discourse practice. This concept's positive connotations make it very useful in the political sphere. It provides a point of departure which, once established, seems self-evident; it is difficult to imagining anyone objecting. *Balance* is a metaphor that tends to disarm opponents (Ashworth 2007: 208). It is interesting to note that the metaphor is also accepted by Swedish critics of video surveillance. Concerns that video surveillance violates people's privacy have been raised by columnists in Swedish newspapers on several occasions. But even this criticism is advanced from the perspective of balancing; the critic differs from the advocate in arguing that the balance has failed (Kjöller 2008; Weiss 2008).

As it turns out, the relativism of privacy is not limited to the process of balancing. A document titled "Balancing Interest According to the Personal Data Act" (Datainspektionen 2009), published by the Swedish Data Inspection Board, states that when balancing interests one must remember that people differ in their definitions of privacy. This subjective view of privacy violation is shared by other judicial actors. The Attorney General (Justitiekanslern) stated in 2008 that protection of privacy is relevant first and foremost for persons for whom surveillance creates discomfort (Lambertz 2008).

The measures taken to increase people's security, including crime prevention, must thus be proportional to the consequent violations of privacy.

This is the essence of balancing; yet in Sweden, the concept's ambiguities allow it to stand for a paradox. The Swedish word for privacy—*integritet*—can mean several different things; thus the opposition between *integritet* (privacy) and security is not always obvious. This paradox can be resolved by shifting the meaning of the word *privacy*. The following contention (made in the Swedish parliament) shows how this might be done: "The most significant violation of privacy (integritet) is to be the victim of a serious crime" (Motion 2007/08 Ju:417). Here, security is defined as integral to privacy: privacy is protected, in part, by being secure from the threat of becoming a victim of crime. Security is thus identified as privacy. The same formulation can be found in a 2006 newspaper article, in which three top dignitaries in the police and judiciary system argue in favour of video surveillance. They claim that victims of crime and terror are subject to violations of privacy (integritet). At the same time, society must have the right to violate people's privacy in order to protect them against crime (Strömberg, et al. 2006). The concept of privacy thus becomes diffuse; nonetheless, the contradiction between privacy and security is dissolved. In 2005, the Swedish public prosecutor (*överåklagaren*) took this argument to its logical extreme when he argued that video surveillance actually increases people's privacy (integritet) because they feel safer and more looked-after (Molin 2005). The radical conclusion must be that a decision not to install cameras would be a violation of, or in any case to detrimental to, privacy. A more nuanced way of formulating the problem is to pit one type of privacy violation against another—the violation caused by surveillance versus the violation of being vulnerable to crime. This is the formulation used by the Ministry of Justice in its guidelines for evaluating the Video Surveillance Act in 2001 (Justitiedepartementet 2001:53). On the whole, the Swedish concept of privacy or personal integrity is not just weak in the sense that it risks being outweighed by other values. Its usefulness in criticism of video surveillance is dubious given that it can also be used in arguments in favour of expanded surveillance without, indeed, doing much damage to the central meaning of the concept.

Poland—Surveillance as Modernisation

In Poland, video surveillance is associated with terms that are strongly positive. The Polish discourse includes the concepts *efficiency, technological development*, and *status*. There is also, however, a (weak) counter-discourse that emphasises both conflicts between different values and the possible violations of *rights*, such as the right to a private sphere, *prywatność*.

Prywatność ('privacy') is defined in the constitution, paragraph 47, as an individual's right to private life, family, honour, and reputation, as well as autonomy in decision making. The Polish constitution accords with the Anglo-Saxon tradition which raises protection of the private sphere to a civil right. The same type of formulation appears in the Polish Civil Code

(Kodeks ciwilny), Article 23, which includes an expanded list of what the private sphere can contain, including the creation of the self and privacy of communication. Together with the Act on Personal Data Protection, these two legal sources are the foundation of the Polish data protection authority's video surveillance policy (Krasinska 2009). The data protection law is also based on a distinct rights perspective. The subject of Article 1:1 of the Act on Personal Data Protection is a person with a right to privacy.

In Polish legal documents, *prywatność* is a traditional, negative right, synonymous with the English word *privacy*. Privacy was originally understood as a liberal freedom, freedom from interference. It is the right to be left alone, to be free from interference and observation by the state. In this sense it is the equivalent of the individual right to private property. This interpretation of the concept has been affirmed by GIODO (Wiewiórowski 2011). However, while privacy defined as non-interference generally has a strong position in public consciousness and is often found in state constitutions, its relevance to the information society is controversial. It is not easy to maintain the right of non-interference in a modern society, transfused as it is by flows of more or less personal data. This is particularly true if *privacy* is seen as an absolute, inviolable right. Indeed, the interpretation of *privacy* in the information age is the subject of a comprehensive international academic debate.

Such ambiguities are clearly revealed in video surveillance. The difficulty of demarcating the private sphere—as a property—in public space is obvious, a fact that has influenced the Polish discourse. The Polish data protection agency refers first and foremost to cases in which cameras were poised so that they filmed people in their homes, for example through a window. The agency has taken action against such violations, understood as a clear infringement of privacy—defined, in turn, as private property. The agency's mandate in such matters has been uncontroversial (Krasinska 2009); a more ambitious video surveillance policy would probably be rejected by the public. Still, this narrow definition does not mean that the idea of right to privacy might not be understood differently and have far-reaching consequences for video surveillance. We will return to this point when we discuss the German discourse.

The view of privacy as a right, absolute or almost absolute, ought to make it more difficult to justify privacy violation on behalf of security or crime prevention (Bennett and Raab 2006: 48). Rights have a different kind of standing; they cannot be relativised as can utilities. To be sure, the Polish discourse also refers to the proportionality principle, in (for instance) critique by the director of the data inspection agency of voice-recording surveillance in schools (a measure initially supported by the Polish government) (Polish Government Decree 2007; Waglowski 2008c). In general, however, the idea of a balance between the protection of the private and other values has not been prominent in Poland. The principle is not written into legal texts to the extent it is in Sweden. As noted above, however, it is unclear to what degree video surveillance actually falls within the framework of something that, in a substantial way, violates the private sphere as it is defined in Poland.

This can be interpreted to mean that video surveillance in Poland is not much associated with the risk of violating the private, despite rigorous judicial protection of the private sphere. The private, thus defined, is only marginally relevant when it comes to regulating surveillance of public places. This framing of the Polish discursive practice naturally reduces opportunities for politicising video surveillance. Critics would have to begin by problematising the basic concept of privacy. This is less likely to occur in a country which has, as a EU member state, quite recently adopted a Western European discourse which often defines the private sphere in a traditional, liberal sense, regarding privacy as a central driving force in social development and, therefore, as something to be protected.[8] The view of the private sphere as private property has become part of post-communist Poland's modernisation project. It is part of a discourse that represents belonging to Europe, as well as Poland's endorsement of market liberalism as an organisational form. For these reasons, the Polish discourse of video surveillance has instead concentrated on 'non-political' issues, such as the 2006 discussion about whether cameras should be permitted in ambulances. In this case, the cameras were removed on the grounds of the patient's right to *intymność* (an intimate sphere) (Pochrzest 2007). *Intymność* is an alternative concept, different from that of privacy, which is very clearly defined as personal property—that is, something that belongs to the individual.

Nonetheless, the fact that civil rights are part of the terminology used to discuss video surveillance opens the way for activism, in the form of movements who see it as their task to sound the alarm when public authorities violate individual rights. As mentioned above, there is a least one tradition of such activism in Poland, although it is not extensive (Panoptykon Fundacja 2011; Waglowski 2011). These activists oppose surveillance in general, but target also video surveillance. Their arguments draw their inspiration from a wider European discourse on human rights; interestingly, they do not ever allude to Poland's past as an authoritarian surveillance state (Svenonius 2011). The criticism in Poland is linked to a national discourse that has delegated communist society to the past; Poland is now a modern, West European country.

Of course, it is not easy to voice critical viewpoints in a country that generally associates video surveillance with something positive. Cameras have a positive symbolic charge in Poland. The argument for expanding video surveillance is linked to crime prevention, in Poland as in Sweden. However, crime prevention in Poland has a somewhat different connotation; cameras fulfil a different role. In the Swedish case, it is primarily the existence of the camera—well identified with signs, so that people know it is there—that will supposedly reduce crime. Such cameras are, as mentioned earlier, commonly unmanned. By contrast, Polish cameras are usually manned. An operator sits and watches the recording in real time; he or she is expected to take action if something suspicious is happening. (One can mention, here, the media storm of 2009, when it was revealed that operators gained a bonus if their reports made it possible to interrupt a crime in progress—which gave them an incentive to do nothing at all until the crime was actually underway; see Gniadek 2009; Machajski 2008.) Nor is there a requirement in Poland that people be informed of on-going video surveillance. There is generally little information about surveillance in areas where cameras are installed.

This is a type of 'situation-directed prevention' different from that used in Sweden. In the Polish discourse the focus is on capturing the perpetrator at the scene of the crime, not deterring the person before the crime is committed. This is why there is no requirement that signs inform the public about the existence of cameras. The logic is as follows: an unmarked camera allows the police to zero in on a dishonest, criminal person; if signs are posted, the dishonest person will go free. This line of thinking assumes that certain people are honest and others are not; there is a clear distinction between 'us' and 'them', the law-abiding and those who break the law. The difference between this, and the Swedish policy of monitoring the each and every member of the general public, is obvious. According to the head of the operational division of the Warsaw video surveillance unit, camera operators are trained in how to recognize, on sight, a person with criminal tendencies or a criminal past—be it a car thief, a drug dealer, or a pickpocket (Janiszemski 2010). A local webpage explains the benefits of the Polish video surveillance system thus: "It is not just the possibility to catch the perpetrator in the act that is very important; by observing life on the street the police can predict dangers and take action. Thanks to this system, information reaches the police immediately, and in this way people committing assault, car thieves, thieves and even minors who are drinking and engaging in hooliganism can be arrested" (Najświeższe informacje lokalne 2007). In addition to a specific view on crime prevention, this passage describes an effective video surveillance system, but it is above all a description of a modern, effective police force that can respond immediately when a crime is taking place. The description might bring to mind an all-powerful police state, but representatives of the Warsaw city security department argue that the camera operators, as civilian employees of local government, should not be seen purely as servants of the state. They are, it is maintained, part of civil society. Camera operators are, thus, civilian guards that should (claim authorities) be viewed as ordinary citizens who have access to technically advanced equipment with which they can alarm the police when necessary (Janiszemski 2010).

Efficiency, defined as the opportunity to intervene in an ongoing crime, is a central argument in Polish video surveillance. The Polish police seem to be convinced of its usefulness for this purpose (Więckowski 2009). Their homepage is regularly updated with extensive reports on arrests made possible by video surveillance (Policja Poland 2011). Meanwhile, the reports presented by the Warsaw city security department show a clear decline in the number of crimes—in areas with cameras, a decrease of 60–70% (Gniadek 2009). We cannot, here, critically evaluate these numbers, although it can be pointed out that the results are extraordinary; nowhere else has video surveillance produced equivalent results. Nor is there lack of criticism in Poland. The effectiveness of video surveillance has been questioned by, among others, researchers at the University of Warsaw (see Waszkiewicz, this volume; Waszkiewicz 2011; Miłosz 2008). The point of statistics of this sort lies, however, not in the exact numbers but in their usefulness in portraying a competent security apparatus that is goal-oriented and rational. This discourse also includes arguments showing how video surveillance images have been used to secure evidence after a crime has been committed, an argument that is also prominent in the Polish policy(see Szrubka, this volume).

One might interpret this attachment to surveillance as an indication that Poles have great faith in technological development—a kind of techno-optimism. However, this is not necessarily the case. Public authorities join private institutions and organisations in presenting video surveillance as an attractive resource, but cameras have taken on a significance that is independent of the immediate uses of the technique. Those who wish can read on the Internet about the ambition to cover all of Warsaw with cameras (Wirtualna Warszawa 2008), an ambition that was, indeed, largely realized by the end of 2009. Proponents of video surveillance emphasize the status that goes along with having cameras; in Poland, camera coverage increases the attraction of a city or institution. Leading bureaucrats in the Warsaw city administration boast that the city's pre-eminent surveillance system compares well with that of foreign cities (Janiszemski 2010). Even critics agree that urban video surveillance is a marketing instrument that signals progress and modernity (Panoptykon 2009). Video surveillance stands for a lifestyle that is associated with the modern metropolis (Szymielewicz 2009). In more small-scale public administrative units, such as schools, video surveillance also contributes to a positive image. Principals note that good surveillance systems influence parents' choice of schools (Dorau 2009). The same processes are observed in choice of housing and neighbourhoods; indeed, Warsaw is a city with a high proportion (by European standards) of well-to-do "gated communities" (Polanska 2011). Video surveillance is part of this concept. The mechanisms behind this logic have been detailed in research by, among others, Koskela (2006: 168), who has shown how so-called city-cams can be used to project a positive picture of a city.

Polish cities commonly advertise video surveillance by posting signs at the city limits. Such signs fulfil an altogether different function than those posted in Sweden, where the object is to inform individuals exactly where and when they are being filmed. The Poles rarely offer this type of information. This is because it does not seem central, or even obvious, that surveillance might be a violation of privacy, as defined in the Polish discourse. Posting signs is meant, rather, to contribute to the positive image of an operation, or city; signs are therefore placed at city borders rather than in the actual places filmed by cameras. The signs are meant to present the city as an attractive place, not to warn individuals who might wish to avoid being filmed.

If video surveillance is to be used in marketing cities, neighbourhoods, and institutions, it is important that the cameras be visible. There were discussions, during the past decade, about whether or not to make civilian guards' video recordings available to the public on the Internet—one way of rendering operations visible. In the Polish city of Wąbrzeźno, for example, there were plans to allow for inter-communicative Internet access to such video recordings. The public, further, was invited to send in its own observations via home computers. The mayor worried, however, that such a system could be used by criminals. This spoke in favour of two systems, one for the public use and one exclusively for the city's organisation of security watch guards (Kozuta 2009). Similar discussions concerning public to access to video recordings have occurred in Warsaw (Waglowski 2008b). Visibility is further promoted by locating camera operators' offices in the same buildings as the offices of other public administrators. The operators work in a space accessible to the public, something that would be unthinkable in, say, Sweden (Wąbrzeźno 2009). In general, it can be said that the protection of material gathered by surveillance is weak in Poland. The length of time for which material can be saved is regulated by practice, not legislation.

Germany—Surveillance and Citizens' Rights

The German discourse on video surveillance is framed in ways that differ from both the Polish and Swedish ones. That which distinguishes it is, above all, its clear connection to a broader discourse on democracy and constitutional rule. *Citizens' rights* and the *right of informational self-determination* are the two central concepts that fill this function. *Security, law and order,* and *consent* are also terms central to the discourse.

To a large extent, the German discourse has the character of a counter-discourse. In the German context this means that the main task is to incorporate video surveillance in a legalistic framework that is compatible with the German self-understanding as a society ruled by law. Characterising this as a counter-discourse is not to say that negative views of video surveillance dominate. Video surveillance has expanded in this country, as it has in other countries, albeit not as explosively as in Sweden and Poland (and—above all—the UK). It is true that there are strong forces in Germany, both in civil society and academic environments, which caution against what they fear is the development of a totalitarian surveillance state. The term *counter-discourse* is used, here, however, to designate the fact that video surveillance in Germany, as well as other types of surveillance, can only gain legitimacy if they take into account views on what is acceptable within the framework of the German Basic Law (*Grundgesetz*). The constitution, and interpretations of the constitution, are the point of departure for discourses about surveillance, including video surveillance. It is against these that positions and actions must be measured. The German legal demand for evidence stands in strong contrast to the Swedish and Polish legal practices. The extensive advisory and negotiation activity in which the states' and the federal data inspection authorities are involved concerning both public institutions and private companies is an important component in this regard. In contrast to Swedish and Polish legal practice where the decision about whether privacy has been violated or not is merely an administrative procedure, the German sets out by defining possible violations in advance; it politicises the issue and demands guarantees that violations *will not* occur.

Germany's modern history is in many ways unique. Its citizens' experience of both Nazism and Communism is impossible to ignore. These experiences include that of a totalitarian state, the German Democratic Republic, which was based on various kinds of surveillance. It is against this background, which includes an awareness of the risks of 'state terrorism', that the debate on video surveillance in Germany must be understood (Töpfer 2004: 6). Criticism of the surveillance state gains sustenance from the country's historical legacy. A metaphor used to describe a totalitarian, controlling state (in terms of its antithesis to the right to informational self-determination) is "citizens of glass" (*Gläserne Bürger*, BfDI 2007b). This concept can be traced to the Constitutional Court's 1983 decision (see below). It evokes a citizen who lacks all privacy, capturing both the individual and societal aspects of the protection of personal rights—that is, citizens' rights. The concept has been developed in subcategories such as *Gläserne Kunde* (BfDI 2005) and *Gläserne Autofahrer* (BfDI 2011b).

The German conceptual apparatus concerning the *private sphere* is more extensive and complicated than that of Sweden or of Poland—the most extensive, perhaps, in the world. The German language has no direct counterpart to the English word *privacy*. It can be said, however, that English's *privacy* is the equivalent of German's *Privatsphäre*. Protection of *the private sphere* includes the rights that belong to the individual in terms of physical privacy, personal correspondence, and the like. As a judicial concept, the *Privatsphäre* is a qualification of the more general concept of *privacy* and includes only a specified part of the rights that are to be protected. In addition to this, the German understanding of privacy includes protection of an intimate sphere (*Intimsphäre*), which alludes, for example, to the autonomy of personal feelings, ideas, and sexual preference. Yet another aspect of the protection of privacy is the so-called *Individualsphäre*; this is first and foremost the right to self-determination (DeSimone 2010: 293ff). The constitutional Basic Law guarantees data protection in Article 1:1 on the right to human dignity and in Article 2:1 on the individual's right to the free development of his or her personality. Together, these formulations come close to a concept of positive freedom: one is guaranteed the right to develop one's own abilities. This goes further than the negative freedom—core to the Anglo-Saxon understanding of privacy—of the right to be left alone.

However, constitutional formulations as such are not the primary influence on practices and discourses of German data protection and video surveillance. More important is the Constitutional Court's interpretation of the right to the free development of one's own personality. This was established in a ruling passed in connection with controversies concerning a census planned for 1983. In this case, the Constitutional Court ruled in favour of the concept *informationelle Selbstbestimmung* (informational self-determination; see Bennett 1992: 76; Kammerer 2008; DeSimone 2010). Since then, this concept has been part of German legal practice. Informational self-determination refers to the individual's right to control over information that is collected on him or her—control over its spread and use. This limits how much individual information the German state can legitimately demand. A lack of control over the collection of information about an individual was described by the Constitutional Court as not only a threat to the individual but a threat to the public at large. Informational self-determination was thus defined as a prerequisite for a democracy based on citizens' freedom to participate in societal activities. In Kammerer's words, "In a society governed by democratic principles and the rule of law, according to the German constitutional court, the free development of a personality is not only a prerequisite for the individual, but also for the common welfare." (Freie persönlichkeitsentfaltung, so des Bundesverfassungsgericht, ist in einer demokratischen rechtsstatlichen Ordning Bedingung nicht nur der individuellen Entfaltungschancen, sonders auch des Gemeinwohls; 2008: 68). Ever since the early 1980s, German discourse has repeatedly returned to the collective, societal benefits that attend the protection of personal information.

This is, then, a right that goes beyond the right covered by the Polish *prywatność*, which is limited to the right to be left alone. Informational self-determination extends the private beyond the protection of an individual's rights. The German understanding is, in fact, closer to a communitarian approach, according to which privacy is one value among others—values whose legitimacy is determined by their social relevance (Etzioni 2007: 1999). However, the German discourse does not have a primarily communitarian origin. The Constitutional Court's 1983 decision does not discuss the right to informational self-determination as something to be balanced against other societal values. It is, rather, seen as an absolute prerequisite for the proper functioning of a democratic society and is thus an essentially liberal right. Regan (1995) has shown that reasoning based on a liberal point of departure can cover the private's societal and collective value. She also argues that information is a source of power, which makes the private—including control over information about oneself—important as a means of holding power accountable and limiting its exercise (Regan 2011). Likewise, the German discussion always centres on the *liberal* constitutional state.

In the German discourse, one speaks about citizens' rights rather than civil rights.[9] We live in a time during which there is an existing threat to the rule of law (Abgeordnetenhaus Berlin 2007), as it was put by a Berlin MP, critical to surveillance, during a 2007 Berlin parliamentary debate. A well-functioning constitutional democratic state depends on the citizen's right to informational self-determination. Given this perspective, the surveillance practice most feared is that exercised over public space by the police—acting, then, as a state authority with a monopoly on violence, representing the state as an instrument of control. The police are not automatically given such power. For some time, the CDU initiated campaigns in various federal states calling for changes to state police laws allowing the police to conduct surveillance of public spaces. Although the Berlin parliament rejected the proposal on several occasions, police surveillance is today permitted in all federal states, including Berlin.

In contrast to the Swedish and Polish data protection legislation, video surveillance is expressly regulated in section 6b:1 of Germany's Federal Data Protection Act. This establishes the circumstances under which it is permissible to conduct video surveillance of spaces to which the public has access. The formulations are general, but the fact that they are included in the data protection law means that video surveillance is seen as an important data protection issue. In Germany, unlike in Poland or Sweden, data protection legislation is given political relevance through the work of the data protection authorities on federal and state level. This work is regularly reported and discussed in parliamentary forums (for example, the operations reports that the authorities present to the state and federal parliaments, in which different aspects of video surveillance are discussed).[10]

The right to informational self-determination clearly affects video surveillance of public space. Here, the threat to citizens' political rights is particularly clear. In contrast to the Anglo-Saxon or Polish concept of privacy, which focuses on self-determination in terms of private space, the right to informational self-determination concerns that which occurs in public, political space. The controversy then sometimes shifts to discussions of which spaces should be considered 'public'. The annual report that Berlin's Commissioner for Data Protection and the Freedom of Information (Berliner Beauftragten für Datenschutz und Informationsfreiheit) presented to the state senate (Berlin's executive governing body; Senat von Berlin) in 2010 makes it clear that this matter is not yet settled. The authority criticised the use of camera equipment in schools, using arguments derived from the legislation on data protection. It argued, among other things, that such measures were not compatible with schools' mandate to instil pupils with the spirit of good citizenship, democracy, peace, and human rights. The senate's response to the report makes it clear that it does not agree with the authority's point of view. School environments are not publicly accessible spaces and therefore fell outside the Data Protection Law's provisions concerning video surveillance (Abgeordnetenhaus Berlin 2010).

If the principle of self-determination over information about one's own person is to function in practice, then people must be informed when such information is being collected. One must, for instance, always be informed when and where video surveillance occurs. Federal German legislation does not mandate signs informing people that they are being videotaped. However, the data protection law calls for appropriate measures to make clear that an area is under surveillance—and to identify who it is doing the surveying. Further, after the intervention of the federal German data protection commissioner, these appropriate measures have, in practice, come to be understood as the posting of signs (Töpfer et al. 2003: 7). Despite the fact that the German legal formulation is weaker than the Swedish, it is clear from its placement in the data protection law that the posting of signs is intended to protect privacy, even if it can also be seen as part of a crime-prevention strategy. The Constitutional Court's ruling also emphasises the importance of an openly visible video surveillance of public spaces as a means of strengthening privacy. The requirement was taken into consideration in a ruling concerning a Bavarian data protection law's compatibility with the constitution, which discussed whether or not it is possible to consent to be monitored (Bundesverfassungsgericht 1 BvR 2368/06 02/23/2007). The possibility of giving or withholding consent is dependent on the availability of knowledge that one might come under surveillance; posting signs, thus, are part of privacy protection. Generally speaking, the demand that there be consent has a strong position in German data protection legislation (Bundesdatenschutsgesetz 1:4a).

Germany, like other countries, hosts an extensive discussion about the efficiency of video surveillance in crime prevention (Christlich-Demokratische Union 2009). In 2000, annual Conference of the Minsters of the Interior (Ständige Konferenz der Innenminister) declared that video surveillance is effective in strengthening a sense of security, preventing crime, and, perhaps, solving crimes that have already occurred (Töpfer 2005: 8). Since then, however, opinion has been divided regarding the actual impact of surveillance. But whereas the Swedish discourse on public video surveillance has focused on crime prevention, the German discussion includes a security concept that goes beyond the more narrow ambition to prevent everyday crime. This is the concept of internal security, *innere Sicherheit*, a term well established in German politics. It concerns different kinds of threats against society, threats whose identity shift over time, depending on shifting constructions of the 'enemy'. The concept originated in the 1970s, in response to the radical activities of left-wing organisations which were seen as threatening vital German social functions. The term has been used, subsequently, in reference to organized crime and terrorism (Kunz 2005: 235ff). It has played a large role in the video surveillance discourse, for video surveillance has been used to counter different kinds of threats against society for the last fifteen to twenty years. The annual Conference of the Ministers of the Interior sanctioned video surveillance as a tool to improve internal security on at least two occasions since 2000 (see Svenonius, this volume).

Recently, however, a distinction between *subjective* and *objective* security (Abgeordnetenhaus Berlin 2003; Die Linke 2009) has received increasing attention (see Svenonius 2011: 193). This distinction is said to lie in the difference between actual security and the feeling of being secure. The concept of subjective security is used to argue that people feel safer when they know that there is video surveillance. This subjective feeling, which exists (it is maintained) regardless of the objective efficiency of video surveillance, has been cited by both proponents and opponents. The CDU argues that a sense of security has become an increasingly important component in individuals' experience of well-being (Christlich-Demokratische Union 2009). Here, video surveillance may make possible a new type of marginalisation. In this case, the enemy would consist of groups who in one way or another fail to fit into a well-organised, modern society, based on the idea of individual success and prosperity. Socially marginalised people tend to be seen as a threat to the secure life of society's well-to-do. The CDU's demand for more law and order fits in here as an argument for increased video surveillance.

The CDU has always referred to law and order in its political promotion of expanded surveillance. Traditional conservative arguments of this sort have been useful in efforts to pass legislation to increase police—and, thereby, state—powers. Law and order arguments may also be complemented by the goal of keeping business operations free from people and groups who cannot afford to consume, and who disrupt a consumption discourse build on social success. Security and cleanliness (*Sauberkeit*) are intimately connected within the image of the city as a potentially threatening environment (Kammerer 2008: 97).

Many German academics characterises this discourse as a neo-liberal security policy applied to urban space (Eick et al. 2007, 9ff). *Neo-liberalisation* (Loader and Walker 2007) refers to the transformation of society according to a model which prioritises consumerism. Neo-liberalism promotes the power of the market, favours privatisation of the public sector, and seeks to build institutions and environments that facilitate these mechanisms' free operation, as anchored in the free choices of rational individuals. Video surveillance is an accepted component of neo-liberal societal architecture. It makes it possible to exclude non-desirable components from the 'clean' urban environments which are the foundation of a consumption society attractive to the middle class. Impersonal technology is a more discrete instrument than human presence, which personifies control. Social control, the usefulness of which is unclear in a neo-liberal context in which morality is individualised, need not be involved.

The critical analysis of neo-liberal development, like the analysis of the security discourse found in German critical criminology (Kunz 2005: 243ff), has spread outside narrow academic environments and into the political sphere. It is interesting that in 2007, the representatives of Berlin's Social Democratic Party supported the right of the police to use video surveillance on the grounds that this protected *citizens'* rights. If police authority were *not* extended, video surveillance would be given over to private security services; and the citizens would be deprived of all insight into the practice (Abgeordnetenhaus Berlin 2007). Opposition to the transfer of power to private security services has also increased among left-wing organisations and MPs critical to surveillance in general. It should be noted, however, that the rhetoric of the Left still focuses on the threat of the totalitarian surveillance state, as manifest in full-coverage video surveillance (Flächendeckende Überwachung; Die Linke 2011).

The Left and the Greens argue that an increased presence of security personnel is preferable to video surveillance. According to these political parties, it is human presence that makes public places more secure (Die Linke 2009; Die Linke 2011; Die Grünen 2011). This applies to all social presence, including various types of guards. It is politically impossible to ignore the issue of security altogether, particularly after the thwarted bombing of a commuter train in Koblenz in the summer of 2006. All political parties have accepted the view that we live in dangerous times. Where they differ is what to do about it. For the left, the demand for an increased human presence represents more than a crime prevention policy; it is an extension of a social analysis that critiques neo-liberal social development—realised in extended video surveillance.

In the political arena, the German discourse is formulated in relation to data protection and the principle of the right to informational self-determination—in contrast to Sweden, where the discourse centres on the proportionality principle. The CDU, for example, clearly takes the critical position and warns of the risk that data protection is invoked as a protection of those who commit crimes, "Datenschutz darf kein Täterschutz sein" (Christlich-Demokratische Union Berlin 2011), while the political Left argues that security policy must leave citizens' rights inviolate. The political arena thus houses antagonistic positions and principles. Again, this is different from Sweden, where the idea of proportionality—with its roots in a bureaucratic discourse practice—also appeals to both positive and negative positions in the political debate.

Naturally, the proportionality principle (Verhältnismäßigkeitsprinzip) is also present in the German discourse. It is one point of departure among others in the judicial formulation of arguments and decisions concerning video surveillance (Bundesarbeitsgerichts AZ: 1 ABR 16/07 08/26/2008). Above all, the concept of balance has been important when passing new legislation that limits individual citizens' rights. Given the German constitution's clear and detailed formulations on the protection of the right to personal privacy and personal development, it has been a delicate process to limit these in the interest of matters that demand the monitoring of citizens. Nonetheless, recent legal practice has shifted the meaning of the constitutional formulations. According to researchers, since as early as the 1970s the interpretation of Articles 1 and 2 of the constitution has increasingly been that the emphasis should be on the responsibility of the state to protect citizens' civil rights—not only as individual rights but also as a collective value. Reformulated in security terminology, the collective value of security has taken on the character of an objective fact based on descriptions of a national threat situation that is difficult to challenge. Increasingly, "facts"—about security threats—tend to be given more weight than rights because, in principle, it is impossible to weigh the two against each other in a meaningful way (Lepsius 2004: 456ff).

The interpretations made by the Constitutional Court in this direction have made it possible to adopt new legislation in the interest of security at both the state and federal levels. Some examples of this are the Law on Organised Crime (OK-Gesetz) of 1992, the Federal Border Police Act (Bundesgrenzschutzgesetz) of 1994, and the Federal Penal Code (Bundeskriminalamtgesetz) of 1998, as well as a number of anti-terrorism laws. These laws concern collection and transfer of technologically collected data about individuals and are therefore more or less applicable to video surveillance. At this level it is not a matter of crime prevention in public spaces or security in a subjective sense—that is, the individual's well-being. Rather, it is about the security of the individual, which is weaved into a security concept that includes the state and society as a whole. However, the German protection of the right to personal development and informational self-determination—which are strong in a comparative perspective—cannot be said to imply any absolute protection (see Töpfer, this volume).

SOCIAL REGIMES OF VIDEO SURVEILLANCE IN SWEDEN, GERMANY, AND POLAND

Table 2.1 is meant to elucidate the three national discourses. We can see how discourse and discourse practice lead to different social regimes—the social structures that frame the conversation on video surveillance. The table summarises the discourses and discursive practices, but it also says something about where each discourse is formed—that is, about who controls it. It is important to keep in mind that we are speaking here of control of the discourse rather than over surveillance itself. A discourse that includes a large portion of restrictiveness, as the German discourse does, might well be compatible with an extensive surveillance expansion.

The discourse on balancing interests concerns weighing different interests and satisfying different needs. It is, in other words, about state-distributed goods and utilities. The state has the ultimate responsibility for

Table 2.1 Video Surveillance as Discourse, Discursive Practice, and Social Regime

	Sweden	Poland	Germany
Discourse	The discourse focuses on balancing interests—the balance metaphor is in focus Pronounced utilitarian framework Crime prevention as public surveillance *Integritet* (privacy) has a weak position—utility rather than a right Posting signs as prevention	The discourse is to a large extent about modernisation Video surveillance as a status symbol Crime prevention as intervention in ongoing crimes *Prywatność*, (privacy)—civil right with little relevance for video surveillance Posting signs as marketing	Citizens' rights and rule of law are central in the discourse Video surveillance is integrated into a larger discourse about the authority of the state Crime prevention and security in focus *Informationelle Selbstbestimmung* (right of informational self-determination) has political relevance Posting signs as privacy protection
Discourse practice	Some political practice, but its content is non-political Weak public practice A bureaucratic practice dominates	Non-existent political practice Weak public practice but with a hint of activist element Uncorrelated, parallel practices	The political, judicial-bureaucratic and public discourses are integrated Strong public practice The judicial practice is central
Social regime	Guardianship; surveillance as distribution by a "well-meaning" state	Fragmentation	Constitutional state as corrective—surveillance requires legitimation

citizens' security and for fighting crime, but is also supposed to guard their personal privacy. One can term the social regime linked to this discourse guardianship. In the Swedish case, guardianship has its anchor in a model of social welfare in which different needs are weighed against each other according to a utilitarian calculus which decentralizes traditional liberal rights as discursive markers.

This discourse characterises the state not as an authoritarian guardian but as benevolent in its relation to its citizens. The state is guided by people's welfare; and it knows what is best for the individual. Situational crime prevention is essentially socially neutral. So is, it would seem, the state's conception of people's propensity to abide by the law; that, also, is distributed impartially. The same logic applies to the state's responsibility to treat everyone equally in measures taken to protect citizens from crime.

Swedish video surveillance is fenced in by specific legislation, but this does not entail much judicial protection against violations of privacy. The interpretation and application of the law is not predetermined; it is, rather, decided in the process of weighing interests against other interests and needs. The balance-of-interests metaphor is fundamental to Swedish discourse. Indeed, it serves to legitimate surveillance by making it politically relevant, yet presents it as a non-controversial issue concerning equitable distribution of different interests. The issue of video surveillance is placed outside traditional ideological differences and other social interest conflicts. Privacy affects us all equally, if it affects any one person individually; and so the utilitarian evaluation of the policy can be left to bureaucrats and experts because the political sphere does not have much to add. The discourse on balancing interests is connected to a discursive practice which gives considerable space to purely administrative and bureaucratic practices. It is relatively difficult for a critical discourse to get a foothold in this discursive environment.

The inclusive character of privacy, *integritet*, which covers both the protection of the private and the security of not becoming a victim of a crime, also disarms criticism against video surveillance. Above all, as we have seen herein, when privacy is conflated with security, it is simply incorporated into the state's traditional areas of responsibilities. The terminological correspondence makes it more difficult, further, to initiate a discussion about what the state can and should do.

Discourses on video surveillance, like those on modernisation, centre on technological development—seen not as a dystopic vision of the future but as an illustration of its potential and possibilities. Technical competence is constructed as progress and well-being, but above all as a measure of modernity. Modernisation, a discourse used in driving for change, often prioritises advanced technology. In Poland, this driving force seems to have had great success, not least because access to technologically advanced video equipment gives the possessor coveted status. Whether on a small or large scale, the issue can be framed in terms of moving Polish society closer to a European identity of progressiveness.

The Polish discourse had failed to reconcile very permissive surveillance practices with judicial regulations regarding protection of the private (*prywatność*). This is due to institutional factors; Poland lacks a coherent institutional structure that can manage surveillance as a policy field. Expansion of video surveillance seems to occur independently of what are, from a privacy perspective, rather promising legal conditions. But this is also caused by the Polish definition of the private, which concentrates on the limits imposed on video surveillance by individual property rights. The private as personal property is linked to the domestic sphere. This makes it difficult to define the type of privacy that might be violated by video surveillance in public places. It does not help, then, that privacy has relatively strong protection as a right. The public practice that draws on a rights tradition for its argument has difficulty asserting itself.

In the German discourse, video surveillance is framed as a question with clear democratic relevance. Opponents of video surveillance have managed to embed the issue in discourses of democracy, the constitutional state, and civil rights. However, forces endorsing expanded video surveillance also frame the question as a matter of the state's role in relationship to the citizens. The concept of internal security reformulates the relationship between the state and the individual, legitimizing limits to individual freedom.

In the German context, it is not possible to separate video surveillance from other questions concerning the surveillance of citizens. Video surveillance is part of a broader discourse in which security is pitted against protection of the private sphere. The state's role is multi-faceted. On the one hand, critics of surveillance portray the state as an agent of control, a threat to the rights of citizens. An alternative view of the state appears in the intentions expressed in the constitution and in the data protection law; here the individual is protected from undue interference and the state is the legitimate protector of citizens' rights. Criticism of video surveillance has, to a large extent, shifted to defending the existing order and existing legislation against interests that would weaken it. There is strong protection against violations of privacy, but this is threatened when privacy is seen as an obstacle to security, either the state's or the individual's. In the discourse on surveillance and citizens' rights, this conflict is clearly formulated.

In the German case, all three discourse practices discussed here—the political, the judicial-bureaucratic, and the public—overlap. Surveillance does not, as in the Swedish case, tend to be first and foremost a bureaucratic question of properly balancing interests. This is because the discourse focuses on questions of a principle nature; they are issues about the organisation of society at a fundamental level. Video surveillance represents something more than a bureaucratic practice.

The practices of posting of signs in the three countries are significant for the discourses. Cameras are not always easy for passers-by to detect. Yet this may be viewed as acceptable in a modernisation discourse. In the Polish discourse, where cameras signal modern status, it is important that everybody knows which actor or institution is holding the cameras. Signs give more information about who owns cameras than about the location of cameras. The situational prevention upon which Swedish practice builds, which depends on unmanned cameras, requires the pre-condition that people know just when and where they are being monitored. Posting signs is thus a prerequisite if the preventive effect is to be achieved—even if signposting is also motivated in terms of protecting privacy. In the German context, arguments of the right to informational self-determination are behind the requirement that signs be posted to inform people about video cameras. Citizens must know when they are being filmed so that they can indirectly consent to (or directly avoid) such monitoring.

An important ingredient in the national discourses is, of course, the different understandings of the private. Sweden's personal privacy has neither a clear political nor judicial content. The Polish *prywatność* differs from the Swedish concept in that the private is a civil right, but also because it is defined as a property—a definition that is not immediately applicable to video surveillance. Behind the German right to informational self-determination is an entirely different logic. It is closely connected to a political practice, and protection of the private in this case becomes a matter of protecting a politically relevant asset—citizens' rights.

NOTES

1. The project published the results as working papers in a series named *On the Threshold to Urban Panopticon? Analysing the Employment of CCTV in European Cities and Assessing its Social and Political Impact.* For the final report see Hempel and Töpfer (2004).
2. As a supplement to the Act on Public Video Surveillance there is legislation covering secret video surveillance, which is regulated in the *Act on Secret Video Surveillance* (Lagen om hemlig kameraövervakning 1995:1506) and the *Act on the Prevention of Certain Particularly Serious Offenses* (Lag om åtgärder för att förhindra vissa särskilt allvarliga brott 2007:979). This chaper does not cover this legislation.
3. For some industries—for example, banks and particular areas in stores—the licensing procedure is not used provided that certain requirements are met. It is sufficient that the county administrative board be informed of the existence of the video camera. An exception is also made for police surveillance for particular purposes and for surveillance that is conducted by particular public authorities—for example, the National Highway Authority (Vägverket).
4. An official government report from 2009 about revising the video surveillance legislation (En Ny Kameraövervakningslag; see Statens offentliga utredningar 2009:87) proposed that general responsibility for video surveillance questions should be transferred from the county boards to the Data Inspection Board. If this were to be adopted, it could have a significant impact on Swedish policy.
5. This said, it should also be noted that extensive civil society criticism has arisen in related questions such as the right of the Defence Ministry's Radio Signalling Unit (Försvarets radioanstalt) to bug telephone calls to and from Sweden. Why these two questions have led to very different responses is a subject to be studied elsewhere.
6. Crime prevention was first introduced as a concept in the legislation in 1990.
7. In an appendix to the report *Ingegritet, Offentlighet, Informationsteckning* (Statens offentliga utredningar 1997:39) which is the written record of the preparatory work of the Personal Data Act, one section is devoted to the question of whether personal privacy is a right. The conclusion is that it is a right, although not an absolute one (appendix 4, 799ff). However, this understanding has not had an impact on the Swedish discourse.
8. The extent to which such a perspective has been adopted in a Polish discourse can naturally be discussed. See Lipiński 2010, which deconstructs three different discursive positions in the Polish debate regarding the view of the EU.
9. Civil rights is an Anglo-Saxon concept that can be linked to a more traditional liberal view of privacy, with its original distinction between the civil society and the political.
10. See, for example, BfDI 2007a, BfDI 2009.

REFERENCES

Bibliography

Ashworth, Andrew. 2007. "Security, Terrorism and the Value of Human Rights." In *Security and Human Rights*, ed. Benjamin J. Goold and Liora Lazarus. Portland, Ore.: Hart, 203–26.

Beck, Ulrich. 1992. *Risk Society: Towards a New Modernity*. London: Sage.

Bennett, Colin J. 1992. *Regulating Privacy: Data Protection and Public Policy in Europe and the United States*. Ithaca, N.Y.: Cornell University Press.

Bennett, Colin J. 2008. *The Privacy Advocates: Resisting the Spread of Surveillance*. Cambridge, Mass.: MIT Press.

Bennett, Colin J., and Charles D. Raab. 2006. *The Governance of Privacy: Policy Instruments in Global Perspective*. Cambridge, Mass.: MIT Press.

Bjereld, Ulf, and Marie Demker. 2009. "Frihet och övervakning." In *I Europamissionens tjänst*, ed. Claes G. Alvstam, Birgitta Jännebring, and Daniel Naurin. Göteborg, Sweden: Centrum för Europaforskning, 241–250.

Björklund, Fredrika. 2011. "Pure Flour in Your Bag: Governmental Rationalities of Camera Surveillance in Sweden." *Information Polity* 16: 355–68.

Björklund, Fredrika. 2012. "The Art of Balancing Utilities: Privacy and Video Surveillance in Sweden." Unpublished manuscript.

Blixt, Madeleine. 2003. *Kameraövervakning i brottsförebyggande syfte*. Report no. 2003:11. Stockholm: Brottsförebyggande rådet.

Borch, Christian. 2005. "Crime Prevention as Totalitarian Biopolitics." *Journal of Scandinavian Studies in Criminology and Crime Prevention* 6: 91–105.

Brottsförebygganderådet. 2007. *Kameraövervakning och brottsprevention*. Report no. 2007:29. Stockholm: Brottsförebyggande rådet.

Cameron, Heather. 2004. "CCTV and (In)dividuation." *Surveillance and Society* 2: 136–44.

Carlsson, Eric. 2009. *Medierad övervakning: En studie av övervakningens betydelser i svensk dagspress*. Umeå, Sweden: Umeå Universitet.

Datainspektionen. 2009. *Intresseavvägning enligt personuppgiftslagen*. Stockholm: Datainspektionen.

DeSimone, Christian. 2010. "Pitting Karlsruhe against Luxembourg? German Data Protection and the Contested Implementation of the EU Data Retention Directive." *German Law Journal* 11: 291–317.

Doyle, Aaron, Randy Lippert, and David Lyon, eds. 2012. *Eyes Everywhere: The Global Growth of Camera Surveillance*. London: Routledge.

Eick, Volker, Jens Sambale, and Eric Töfper, eds. 2007. *Kontrollierte Urbanität zur Neoliberalisierung Städtischer Sicherheitspolitik*. Wetzlar, Germany: Transcript-Verlag.

Etzioni, Amitai. 1999. *The Limits of Privacy*. New York: Basic Books.

Etzioni, Amitai. 2007. "Are New Technologies the Enemy of Privacy?" *Knowledge, Technology and Policy* 20: 115–19.

Fairclough, Norman. 2008. *Discourse and Social Change*. Cambridge: Polity Press.

Flaherty, David H. 1989. *Protecting Privacy in Surveillance Societies: The Federal Republic of Germany, Sweden, France, Canada, and the United States*. Chapel Hill: University of North Carolina Press.

Fussey, Pete. 2008. "Beyond Liberty, beyond Security: The Politics of Public Surveillance." *British Politics* 3: 120–35.

Garland, David. 2000. "Ideas, Institutions and Situational Crime Prevention." In *Ethical and Social Perspectives on Situational Crime Prevention*, ed. Andrew von Hirsch, David Garland, and Alison Wakefield. Oxford: Hart.

Gilliom, John. 2001. *Overseers of the Poor: Surveillance, Resistance, and the Limits of Privacy*. Chicago: University of Chicago Press.
Gras, Marianne, L. 2004. "The Legal Regulation of CCTV in Europe." *Surveillance and Society* 2: 216–29.
Gräslund, Göran. 2005. "Likheten inför lagen ur spel av kameror för övervakning. Datainspektionens chef vill ha övergripande ansvar för att kameralagen tillämpas lika över hela landet." *Dagens Nyheter*, February 12.
Gräslund, Göran. 2008. "Kameraövervakningen på skolor är olaglig. Datainspektionen sätter stopp för övervakningskameror på sju granskade skolor: Beslutet är vägledande för alla skolor i landet." *Dagens Nyheter*, October 2.
Haggerty, Kevin D., and Richard V. Ericson. 2000. "The Surveillant Assemblage." *British Journal of Sociology* 51: 605–22.
Hempel, Leon, and Eric Töpfer. 2002. *Working Paper No. 1: Inception Report*. Urbaneye Project. Berlin: Zentrum Technik und Gesellschaft, Technische Universität Berlin.
Hempel, Leon, and Eric Töpfer. 2004. *Working Paper No. 15: CCTV in Europe. Final Report*. Urbaneye Project. Berlin: Zentrum Technik und Gesellschaft, Technische Universität Berlin.
Hempel, Leon, and Eric Töpfer. 2009. "The Surveillance Consensus: Reviewing the Politics of CCTV in Three European Countries." *European Journal of Criminology* 6: 157–77.
Hier, Sean P. 2003. "Probing the Surveillant Assemblage: On the Dialectics of Surveillance Practices as Processes of Social Control." *Surveillance and Society* 1: 399–411.
Jacobsson, Kerstin, ed. 2010. *Känslan för det allmänna. Medborgarnas relation till staten och varandra*. Umeå, Sweden: Boréa.
Kammerer, Dieter. 2008. *Bilder der Überwachung*. Frankfurt am Main: Suhrkamp Verlag.
Kilian, Wolfgang. 2008. "Germany." In *Global Privacy and Protection: The First Generation*, ed. James B. Rule and Graham Greenleaf. Cheltenham, England: Edward Elgar, 80–106.
Kjöller, Hanne. 2008. "Hur kunde något så dyrbart hanteras så billigt." *Dagens Nyheter*, January 16.
Koskela, Hille. 2006. "'The Other Side of Surveillance': Webcams, Power and Agency." In *Theorizing Surveillance: The Panopticon and Beyond*, ed. David Lyon. Cullompton, Devon, England: Willan, 163–181.
Kunz, Thomas. 2005. *Der Sicherheitsdiskurs. Die Innere Sicherheitspolitik und ihre Kritik*. Bielefeld, Germany: Transcript-Verlag.
Laclau, Ernesto, and Chantal Mouffe. 2008. *Hegemonin och den socialistiska strategin*. Göteborg, Sweden: Vertigo.
Lambertz, Göran. 2008. "Lagen tillåter inte dagens kameraövervakning." *Dagens Nyheter*, November 29.
Lazar, Michelle. M., ed. 2005. *Feminist Critical Discourse Analysis: Gender, Power and Ideology in Discourse*. New York: Palgrave Macmillan.
Lepsius, Oliver. 2004. "Liberty, Security and Terrorism: The Legal Position in Germany." *German Law Journal* 5(5): 435–60.
Liberatore, Angela. 2007. "Balancing Security and Democracy, and the Role of Expertise: Biometrics Politics in the European Union." *European Journal on Criminal Policy and Research* 13(1–2): 109–37.
Lipiński, Artur. 2010. *Europe as a Symbolic Resource: On the Discursive Space of Political Struggles in Poland*. KFG, the Transformative Power of Europe. Working Paper no. 10. Berlin: Frie Universität Berlin.
Loader, Ian, and Neil Walker. 2007. *Civilizing Security*. Cambridge: Cambridge University Press.

Locke, Terry, ed. 2004. *Critical Discourse Analysis*. London: Continuum.
Lomell, Heidi Mork. 2004. "Targeting the Unwanted: Video Surveillance and Categorical Exclusion in Oslo, Norway." *Surveillance and Society* 2: 346–60.
Luhmann, Niklas. 1993. *Risk: A Sociological Theory*. New York: De Gruyter.
Lyon, David, ed. 2003. *Surveillance and Social Sorting: Privacy, Risk and Digital Discrimination*. London: Routledge.
Marx, Gary T. 2002. "What's New about the 'New Surveillance'? Classifying for Change and Continuity." *Surveillance and Society* 1: 9–29.
McCahill, Michael, and Clive Norris. 2003. *Working Paper No. 10: CCTV Systems in London. Their Structures and Practices*. Urbaneye Project. Berlin: Zentrum Technik und Gesellschaft, Technische Universität Berlin.
Molin, Kari. 2005. "Fyra frågor till Sven-Erik Alhem: Låt gå mentaliteten är det värsta som finns." *Dagens Nyheter*, August 19.
Mouffe, Chantal. 2005. *On the Political*. London: Routledge.
Newman, Abraham L. 2008. *Protectors of Privacy: Regulating Personal Data in the Global Economy*. Ithaca, N.Y.: Cornell University Press.
Norris, Clive, and Gary Armstrong. 1999. *The Maximum Surveillance Society: The Rise of CCTV*. New York: Berg.
Norris, Clive, Mike McCahill, and David Wood. 2004. "The Growth of CCTV: A Global Perspective on the International Diffusion of Video Surveillance in Publicly Accessible Space." *Surveillance and Society* 2(2–3): 110–35.
Polanska, Dominika. 2011. "The Emergence of Enclaves of Wealth and Poverty: A Sociological Study of Residential Differentiation in Post-Communist Poland." PhD diss., Stockholm University.
Raab, Charles D. 1999. From Balancing to Steering: New Directions for Data Protection. In *Visions of Privacy: Policy Choices for the Digital Age*, ed. C. J. Bennett and R. Grant. Toronto: University of Toronto Press, 68–93.
Regan, Priscilla M. 1995. *Legislating Privacy: Technology, Social Values, and Public Policy*. Chapel Hill: University of North Carolina Press.
Regan, Priscilla M. 2011. "Response to Bennett: Also in Defense of Privacy." *Surveillance and Society* 8: 497–99.
Rule, James, B. 2007. *Privacy in the Peril: How We Are Sacrificing a Fundamental Right in the Exchange for Security and Convenience*. Oxford: Oxford University Press.
Rule, James B. 2009. "The Limits of Privacy Protection." In *New Directions in Surveillance and Privacy*, ed. Benjamin J. Goold and Daniel Neyland. Cullompton, Devon, England: Willan, 3–17.
Rule, James B., and Graham Greenleaf, eds. 2008. *Global Privacy Protection: The First Generation*. Cheltenham, England: Edward Elgar.
Statens offentliga utredningar. 1974. *Fotografering och Integritet*. Report no. 1974:85. Stockholm: Fritzes.
Statens offentliga utredningar. 1997. *Integritet Offentlighet Informationsteknik*. Report no. 1997:39. Stockholm: Fritzes.
Statens offentliga utredningar. 2002. *Allmän kameraövervakning*. Report no. 2002:110. Stockholm: Fritzes.
Statens offentliga utredningar. 2009. *En ny kameraövervakningslag*. Report no. 2009:87. Stockholm: Fritzes.
Strömberg, Stefan, Fredrik Wersäll and Klas Bergenstrand. 2006. "Sluta måla upp falska bilder av övervakarstaten. Riksåklagaren, rikspolischefen och säpochefen: Vi behöver moderna verktyg för att bekämpa den grova brottsligheten." *Dagens Nyheter*, February 11.
Svenonius, Ola. 2011. *Sensitising Urban Transport Security: Surveillance and Policing in Berlin, Stockholm and Warsaw*. Södertörn Doctoral Dissertations 61. Stockholm: Södertörn University/Stockholm University.

Szekely, Ivan. 2008. "Hungary." In *Global Privacy and Protection: The First Generation*, ed. James B. Rule and Graham Greenleaf, 174–206. Cheltenham, England: Edward Elgar.
Szymielewicz, Katarzyna. 2009. "Miastoobiekt monitorowany?" *Kultura Miasta* 5: 94–101.
Söderlind, Åsa. 2009. *Personlig integritet som informationspolitik. Debatt och diskussion i samband med tillkomsten av Datalag (1973:289).* Borås, Sweden: Valfrids.
Tham, Henrik. 2002. "Law and Order as a Leftist Project?" *Punishment and Society* 3(3): 409–26.
Töpfer, Eric. 2004. *Das Urbaneye-projekt: Videoüberwachung im europäischen Vergleich. Gemeinsame Trends, nationale Unterschiede, Probleme und Perspective. Vortrag für den Workshop "Videoüberwachung, Theorie ind Praxis."* Hamburg: Institut für Kriminologische Sozialforschung der Universität Hamburg.
Töpfer, Eric. 2005. "Die polizeiliche Videoüberwachung der öffentlichen Raums: Entwicklungen und Perspektiven." *Datenschutz Nachrichten* 28(2): 5–9.
Töpfer, Eric, Leon Hempel, and Heather Cameron. 2003. *Working Paper No. 8: Watching the Bear. Networks and Islands of Visual Surveillance in Berlin.* Urbaneye Project. Berlin: Zentrum Technik und Gesellschaft, Technische Universität Berlin.
Waples, Sam, Martin Gill, and Peter Fisher. 2009. "Does CCTV Displace Crime?" *Criminology and Criminal Justice* 9: 207–24.
Waszkiewicz, Paweł. 2011. "Wielki Brat Rok 2010." *Systemy monitoring wizyjnego—aaspekty kryminalistyczne, kryminologiczne I prawne.* Warsaw: Wolters Kluwer Polska.
Webster, William R. 2004. "The Diffusion, Regulation and Governance of Closed-Circuit Television in the UK." *Surveillance and Society* 2: 230–50.
Weiss, Lars. 2008. "Men kamerorna då? Lars Weiss om en växande övervakningsindustri." *Dagens Nyheter*, December 9.
Weiss, Gilbert, and Ruth Wodak, eds. 2003. *Critical Discourse Analysis: Theory and Interdisciplinarity.* Basingstoke, England: Palgrave Macmillan.
Wiewiórowski, Wojciech, R. 2011. "Prawo do prywatności." *Przegląd Komunalny* 11: 72.
Zedner, Lucia. 2007. "Seeking Security by Eroding Rights: The Side-stepping of Due Process." In *Security and Human Rights*, ed. Benjamin J. Goold and Liora Lazarus. Portland, Ore.: Hart, 277–303.

Parliamentary Records

Abgeordnetenhaus Berlin. 2003. 33. Sitzung vom 26. Juni.
Abgeordnetenhaus Berlin. 2007. 21. Sitzung vom 22 November.
Abgeordnetenhaus Berlin. 2010. Drucksache 16/3377. *Stellungnahme des Senats zum Bericht des Berliner Beauftragten für Datenschutz und Informationsfreiheit für das Jahr 2009.*
Christlich-Demokratische Union. 2003. Abgeordnetenhaus Berlin. Antrag der Fraktion der CDU Drucksache 15/1827.
Christlich-Demokratische Union. 2009. Abgeordnetenhaus Berlin. Antrag der Fraktion der CDU Drucksache 16/2266 .
Interpellation 2007/08:383, Social Democratic Party (Swedish parliament).
Justitieutskottets betänkande 1997/98:JuU14, reservation (Swedish parliament).
Motion 1997/98:Ju20, Liberal Party (Swedish parliament).
Motion 1997/98:Ju21, Conservative Party (Swedish parliament).
Motion 1997/98:Ju22, Left Party (Swedish parliament).

Motion 1997/98:Ju910, Center Party (Swedish parliament).
Motion 2005/06:Ju273, Christian Democratic Party (Swedish parliament).
Motion 2007/08:Ju417, Social Democratic Party (Swedish parliament).
Motion 2009/10:Ju348, Conservative Party (Swedish parliament).
Motion 2009/10:Ju405, Social Democratic Party (Swedish parliament).
Polish Government Decree. 2007. 09/06/2007 (nr 1155, Dz.U. z r. Nr 163).
Regeringsproposition 1975/76:194 (Swedish parliament)
Regeringsproposition 1989/90:119 (Swedish parliament)
Regeringsproposition 1997/98:64 (Swedish parliament)
Riksdagsprotokoll 1997/98:86 (Swedish parliament)
Riksdagsprotokoll 2006/07:132 (Swedish parliament)
Riksdagsprotokoll 2006/07:1463 (Swedish parliament)
Riksdagsprotokoll 2007/08:69 (Swedish parliament)
Sozialdemokratische Partei Deutschlands. 2007. Abgeordnetenhaus von Berlin 21. Sitzung vom 22. November.

Interviews

Gniadek, Jacek. 2009. Director of the Unit for Video Surveillance. Interview by Fredrika Björklund, Elfar Loftsson, and Wojciech Szrubka. Interview notes. Warsaw, Biuro Bezpieczeństwa i Zarządzania Kryzysowego. December 16, 2009.
Krasinska, Monika. 2009. Director for Jurisprudence, Legislation, and Complaints at the Polish Data Protection Agency. Interview by Fredrika Björklund, Elfar Loftsson, and Wojciech Szrubka. Interview notes. Warsaw, Offices of Generalny Inspektor Ochrony Danych Osobowych. December 18, 2009.
Więckowski, M. 2009. Head of Police Surveillance Unit. Interview by Ola Svenonius and Mihal Bron. Interview notes. Warsaw, Komisariat policiji, section 1. December 18, 2009.
Panoptykon. 2009. Head of Panoptykon. Interview by Fredrika Björklund, Elfar Loftsson, Ola Svenonius and Wojciech Szrubka. Interview notes. Warsaw, Panoptykon office. December 16, 2009.
Dorau, Ewa. 2009. Rector at School No. 3, Interview by Fredrika Björklund, Elfar Loftsson, and Wojciech Szrubka. Interview notes.. Wąbrzeźno, rector's office. May 6, 2009.
Kozuta, Bogdan. 2009. Mayor of Wąbrzeźno. Interview by Fredrika Björklund, Elfar Loftsson, and Wojciech Szrubka. Interview notes. Wąbrzeźno, mayor's office. May 6 2009.
Wąbrzeźno. 2009. Visit to the Operative Unit for Video Surveillance in Wąbrzeźno. May 5, 2009.

Other Public Documents and Websites

BfDI 2005. Verbraucherpolitik in der digitalen Welt—Der Gläserne Kunde. Der Bundesbeauftragte für den Datenschutz, Pressemitteilung, 04/21/2005.
BfDI 2007a. Tätigkeitsbericht Der Bundesbeauftragte für den Datenschutz und die Informationsfreiheit 2005–2006.
BfDI 2007b. Tätigkeitsbericht Der Bundesbeauftragte für den Datenschutz und die Informationsfreiheit 2005–2006, Pressemitteilung, 04/24/2007.
BfDI 2007c. Bundesbeuftrager für den Datenschutz und die Informationsfreiheit, Pressemitteilung 06/01/2007.
BfDI 2009. Tätigkeitsbericht Der Bundesbeauftragte für den Datenschutz und die Informationsfreiheit 2007–2008.

BfDI. 2010. Bundesbeuftragter für den Datenschutz und die Informationsfreiheit, Pressemitteilung 08/25/2010.
BfDI. 2011a. Bundesdatenschutzbeauftragter stellt 23. Tätigkeitsbericht 2009–2010 vor, Pressemitteilung 04/12/2011.
BfDI. 2011b. Für Elektromobilität, aber gegen gläserne Autofahrer! Der Bundesbeauftragte für den Datenschutz, Pressemitteilung, 05/16/2011.
Bundesarbeitsgerichts AZ: 1 ABR 16/07, 08/26/2008.
Bundesverfassungsgericht 1 BvR 2368/06, 02/23/2007.
Bundesverfassungsgericht 2 BvR 1447/10, 08/12/2010.
Central and Eastern Europe Data Protection Authorities. 2011. Home page. Available at http://www.ceecprivacy.org.
Christlich-Demokratische Union Berlin. 2011. "Antworten der CDU Berlin." Wahlprüfsteine. Available at http://berlin.humanistische-union.de/index.php?id=2698.
Datainspektionen. 2008a. *Beslut efter tillsyn enligt personuppgiftslagen* (1998:204) Dnr 742–2008.
Datainspektionen. 2009. *Informerar: Intresseavvägning enligt personuppgiftslagen*. Brochure.
Datainspektionen. 2011a. Home page. Available at http://www.datainspektionen.se.
Datainspektionen. 2011b. "Kameraövervakning." Available at http://www.datainspektionen.se/lagar-och-regler/personuppgiftslagen/kameraovervakning/.
European Digital Rights. 2008. Home page. Available at http://www.edri.org.
Die Grünen. 2011. "Antworten von Bündnis '90/Die Grünen Berlin." Wahlprüfsteine. Available at http://berlin.humanistische-union.de/themen/wahl_2011/wahlpruefsteine/buendnis_90die_gruenen.
Högsta förvaltningsdomstolen 4459–10 01/31/2011 (Swedish Supreme Administrative Court, superseded Regeringsrätten as supreme administrative court in January 2011).
Janiszemski, Bartosz. 2010. "Warszawiaku, oni cię znają z widzenia." *Newsweek Poland*. Accessed 21 March 2010 at http://spoleczenstwo.newsweek.pl/warszawiaku—oni-cie-znaja-z-widzenia,55518,1,1.html.
Justitiedepartmentet, Instruktion 2001:53 (Swedish Government).
Justitiedepartmentet, Instruktion 2008:22 (Swedish Government).
Krajewski, Łukasz. 2007. "Monitoring jest, ale regulacji wciąż nie ma." *Gazeta Wyborcza*. Accessed 25 September 2007 at http://miasta.gazeta.pl/warszawa/1,95192,4521471.html.
Die Linke. 2009. Landesverband Berlin, "Videoüberwachung." Available at http://www.die-linke-berlin.de/wahlergebnisse.
Die Linke. 2011. *Videoüberwachung zurückdrängen*, Pressemitteilung 02/10/2011 (J. Korte).
Machajski, Piotr. 2008. "Nie czekamt na gwałt—zapewniają urzędnicy." *Gazeta Wyborcza*. Accessed 19 December 2008 at http://wyborcza.pl/1,94898,6086247.html.
Miłosz, Maciej. 2008. "Monitoring nie zwiększa bezpieczeństwa?" *Życie Warszawy*. Accessed 5 November 2008 at http://www.zw.com.pl/artykul/2,303626_Monitoring_nie_zwieksza_bezpieczenstwa__.html.
Najświeższe informacje lokalne. 2007. "System Monitoringu Pomaga Stołecznym Policjantom." Available at http://warszawa.naszemiasto.pl/archiwum/1409700,system-monitoringu-pomaga-stolecznym-policjantom,id,t.html.
Panoptykon Fundacja. 2011. Home page. Available at http://www.panoptykon.org.
Pochrzest, Agnieszka. 2007. "Kamery Już Nie Nagrywają Chorych." Accessed 5 April 2007 at http://warszawa.gazeta.pl/warszawa/1,86775,4042640.html.

Policja Poland. 2011. Home page. Available at http://www.policja.pl.
Regeringsrätten 6572–98 09/29/2000 (Swedish Supreme Administrative Court).
Regeringsrätten 1450–01 09/26/2001 (Swedish Supreme Administrative Court).
Regeringsrätten 5767–02 03/09/2004 (Swedish Supreme Administrative Court).
Regeringsrätten 6462–05 02/02/2009 (Swedish Supreme Administrative Court).
Regeringsrätten 7873–08 02/11/2010 (Swedish Supreme Administrative Court).
Waglowski, Piotr VaGla. 2008a. "W jaki sposób zdjęcia z monitoringu trafiły do mediów?" Accessed 27 May 2008 at http://www.prawo.vagla.pl/node/7893.
Waglowski, Piotr VaGla. 2008b. "A właściwie, to na jakich zasadach działa monitoring miejski w Warszawie?." Accessed 28 May 2008 at http://www.prawo.vagla.pl/node/7894.
Waglowski, Piotr VaGla. 2008c. "Monitoring wizyjny w szkołach." Accessed 22 May 2008 at http://www.prawo.vagla.pl/node/7885.
Waglowski, Piotr VaGla. 2011. Home page. Available at http://www.prawo.vagla.pl.
Wirtualna Warszawa. 2008. "Prawie 600 kamer monitoring w Warszawie." Available at http://www.wirtualna-warszawa.pl/w/080405/prawie-600-kamer-monitoringu-w-warszawie.html.

3 Video Surveillance in a Historical Perspective

Ola Svenonius

INTRODUCTION

Surveillance practices in public spaces are often seen in relation to issues of democratic legitimacy. This is particularly true for video surveillance because of its symbolic character and the high significance given to observation and supervision in Western democratic thought, as recently pointed out by Doyle et al. (2012a: 5). Witnessing the burgeoning of surveillance practices across the globe, critical scholars often argue that there is a discrepancy between the rapid spreading of (especially video) surveillance, and the actual advantages of these applied technologies. Given the danger of a 'surveillance society' looming around the corner, and the relative lack of evidence validating its benefits, video surveillance is often treated particularly sceptically by critical scholars and privacy advocates. The United Kingdom is typically seen as the most 'advanced', or 'promiscuous', video surveillance nation, but recent academic work—including the contributions in this volume—show that the gap is being closed by other countries (Hempel and Töpfer 2004; Norris 2012; Svenonius 2011, 2012; Waszkiewicz 2011). Not only are video surveillance practices expanding rapidly in Europe and North America, surveillance technology is increasingly being used in China to monitor the public and 'maintain order' (Doyle et al. 2012a: 3ff). The Chinese authorities' use of video surveillance embodies much of what European and North American critics fear, with its oppressive dimension of technologically mediated social control. In light of such a global comparison, and taking into account the European data protection legislation, there seems to exist a 'European approach' to surveillance in many respects (see Jonason, this volume). While this is to some degree true, the countries included in this study were selected because of the big differences they display in their cultural and legal understandings of surveillance. In broad terms, one might say that (video) surveillance in Germany is traditionally highly controversial, while in Poland and Sweden such practices have been regarded as perfectly normal. That said, German and Swedish surveillance is thoroughly regulated, whilst in Poland surveillance is virtually unregulated. In Poland, video surveillance is rather viewed as a sign of modernisation and progress, as Björklund's and Szrubka's chapters in this volume argue. Historical and institutional factors surely influence this perception, and in this section we shall take a closer look at the former.

The question that I address in this chapter is how to understand such differences between Germany, Poland, and Sweden from a historical perspective. Below, I advance three fairly simple historical arguments about video surveillance in these countries. I argue that the current strict regulation in Germany is a result of the social and political turmoil of the 1970s, during which German society became highly polarised. In Poland, on the other hand, the loose regulation (or complete lack thereof) is connected to the high fear of crime that Polish citizens felt after 1989. Finally, in Sweden, the roots of the regulation of video surveillance can be traced back to a fairly short period of concern about privacy in the early 1970s—this in a society that is traditionally generally uncritical towards (state) surveillance. In this chapter I aim to develop these arguments by means of a comparison of surveillance systems generally and, by extension, video surveillance. In the three countries included in this study, we find very different approaches that are both surprising and analytically significant.

THEORETICAL PERSPECTIVE AND CHAPTER OUTLINE

The chapter focuses on key moments in the recent political history of (mainly West) Germany, Poland, and Sweden since around 1970. This time frame was chosen because it was in the early 1970s that information technologies began to spread more widely and many of the discussions about what later became known in positive terms as the 'information society' began. Based on a survey of existing literature, this chapter identifies the key events in the area of general surveillance systems and connects them with the regulation of *video* surveillance. The chapter therefore focuses on the historical contingencies of the present-day organisation of video surveillance in all three countries. In order to achieve this goal, I utilise Mouffe's notion of 'the political' as an analytical heuristic to sort through the material and produce a coherent historical narrative (see further discussion below). In order to display historical changes in an accessible way, I discuss surveillance politics in two periods: the 1970s and '80s, and the 1990s to the present. Each section contains a country-specific description and a comparative summary. In the concluding section I discuss the results from the perspective of resistance and national differences in attitudes towards video surveillance. Immediately following I briefly discuss the notion of 'the political'.

Mouffe considers 'the political' to not be limited to specific institutions, but as a field of contestation in which societal norms are established in a constructive, albeit conflict-ridden process. This conflict about what is righteous, fair, acceptable, or otherwise normatively correct is what I take to be 'the political' (Mouffe 1999, 2007). The way that any issue is either made into a political topic or remains depoliticised is at the core of what

I understand to be the political process. This understanding is based on post-structuralist discourse theory, and is a simplified version of conflict theory (Glynos and Howarth 2007: 144ff; Joas and Knöbl 2009: 186ff). The underlying assumption guiding the analysis is that there are several mutually exclusive interests in any given society that struggle for specific institutional arrangements to prevail. While none of these are naturally given, 'the political' in general is the site for struggle about morality and legitimacy of political ideologies. By studying political concepts, one may open the door to a larger political problematic that can be studied comparatively. The benefit of using Mouffe's theory is that it highlights the historical genealogies of regulation and makes it possible to identify the contingency of today's politics. Thus, in any given situation, there is a political reality—an event that somehow creates a political reaction in society. This reaction will be affected by how the event was understood, or perhaps 'framed'. The approach used in this text is therefore not guided by a specific hypothesis or theoretical assumption, but by a desire to describe and understand political history from the perspective of Mouffe's notion of 'the political'. It is a relatively thin theoretical construction designed to let history 'speak as much for itself' as possible. Hence this chapter also serves as a historical background for the other chapters in this volume. Below, I set out to first describe the early period of the politics of information technology and video surveillance, and afterwards proceed to more recent times.

SURVEILLANCE IN THE 1970S AND '80S

The 1970s and '80s were turbulent years. In West Germany, the activities of the terrorist group Red Army Faction (Rote Armee Fraktion, RAF) made new modes of policing and surveillance necessary, while in Sweden the Social Democratic Workers' Party (Socialdemokratiska arbetareparti, SAP) faced several crises arising from the Swedish secret police's illegal records, which had been uncovered by the end of the 1960s. In both West Germany and Sweden, the regulation of video surveillance and data protection in general was characterised by a fear of computerised authoritarianism. In Poland and East Germany, societal turbulence took on entirely different dimensions of course, with one quarter of the whole population being Solidarity (Solidarnosc) members in Poland in 1981, and with the many demonstrations that took place in East Germany towards the end of the 1980s. Below I describe these developments in turn, and towards the end of the section discuss the cases comparatively.

Sweden: 'Personal Integrity' and Computerisation

The history of surveillance and security in Sweden is characterised by a strong regulatory practice formed from a political culture that was relatively insensitive to the privacy interests of the citizens. Swedish society is traditionally characterised by a high degree of political consensus, particularly around the design of the welfare state. A sign of this consensus is the dominance of the SAP, which ruled the country for forty-four consecutive years between 1932 and 1976. This period was one of massive state expansion, in which most of the major welfare state institutions were created. The Swedish model—corporatism, a 'mixed economy' and universal welfare benefits—enjoyed unquestioned legitimacy at least until 1980, and even then questioning the fundamental principles was unthinkable. At the end of this period a number of things happened in Europe, and especially in Sweden, that changed this stability.

The 1960s and '70s witnessed several minor scandals involving political surveillance of left-wing organisations. This culminated in the so-called IB affair. The IB was a secret intelligence agency organised under the regular national secret service (Rikspolisstyrelsen/Säkerhetsavdelningen, RPS/Säk). By the end of the 1960s, after considerable media attention regarding secret intelligence registers, the government instituted a ban on political surveillance, forbidding the RPS/Säk to register citizens' political views. Secretly, however, the government had instructed the RPS/Säk to closely monitor a range of left-wing and socialist organisations (Statens offentliga utredningar 2007: 503ff). The police used quite elaborate techniques to perform their task, including all forms of audio and video surveillance, tapping phones, manual surveillance, and using civilian informers and students to infiltrate organisations. Many techniques resembled those from authoritarian socialist regimes, albeit in milder forms: involvement in a 'marked' organisation could result in intensive secret surveillance, 'security hearings' with the secret service, and exclusion from employment even many years after the initial registration.[1] The information in these political registers was routinely used to exclude left-wing sympathisers from public positions.[2] The resulting sceptical attitude towards the secret services continued well into the 1990s, and finally ended with the publication of a report in which the irrational and habitual surveillance practices were discussed (Statens offentliga utredningar 2002). The over-zealous surveillance practices were exposed and published in the media in 1973, and this scandal effected a minor dislocation in Sweden. The trust in the state, which indirectly meant the SAP, was low, and it worsened in another scandal in 1975 when it turned out that the secret services were still deploying infiltrators, this time in a hospital in Gothenburg (Statens offentliga utredningar 2007: 503). The SAP lost the 1976 and 1979 elections and, together with the economic crises of the 1970s and the massive critique against Keynesian economics that followed, the political landscape went into a transformation which affected not only welfare institutions but crime policy and Swedish policing as well.

By that time, the general computerisation of the public sector had also begun—Sweden was an 'early adopter' in this area (Ilshammar 2002: 121; Bennett 1992: 62)—and with the growing realisation that computers could be used for surveillance, a computer-critical discourse developed in the late 1960s. This fear was realised in the 1970 population census, which triggered an intense debate. The methods used to gather the statistics were perceived as highly intrusive and this, together with the fear of computerisation, resulted in a crisis in confidence not only for the responsible agency Statistics Sweden, but for the SAP in general. Ilshammar (2002: 122ff) shows how the attitude of the parliament changed from optimistic about computers to highly critical. About the attitudes in wider society, he writes, "From being considered to be an efficiency-boosting tool in the expanding welfare state's service, during the late 1960s the computers more and more came to be understood as a surveillance technique and symbol of a large-scale and inhuman control society" (Ilshammar 2002: 146; my translation). The debate focused on the computerisation of the state, resulting in 'inhuman control', 'elite rule', and even biopolitics (genetic engineering and cloning; see Statens offentliga utredningar 2007: 499ff). The emotional response to the unknown—the computer, or 'math-machine'—fuelled positions of anti-statism with emotional content. Only in the 1990s can we see a change from this frightening fantasy of the totalitarian computerised state into a utopian imaginary of technology as modernisation (the 'information super-highway' and the 'information society').

The dystopian visions transformed the computer issue from being a mainly left-wing matter to being a powerful critique of state surveillance. The conflict resulted in a parliamentary commission that produced a draft act on computerised registers—the so-called Data Act (Datalagen, SFS 1973:289) which was approved by the parliament in 1973. The act introduced a general licensing and permission system that subsumed all registers into a new Data Inspection Board (Datainspektionen, or DI; see Bennett 1992: 60ff, 161ff). Ilshammar considers the act to be a compromise between, on the one hand, different political interests in the Swedish administration (such as Statistics Sweden's many registers and habit of selling personal data for advertisements) and, on the other hand, the principle of transparency in the public sector that grants access to all public documents (*offentlighetsprincipen*) and the privacy interests of the population, which is called 'personal integrity' (*personlig integritet*) in Swedish. The conflict around the technocratic Statistics Sweden during the 1970 population census created the conditions for a more general regulation of data protection in Sweden. In fact Sweden's Data Act was the first national data protection legislation in the world, to be closely followed by Germany and the United States. However, the sensitivity to issues of surveillance that grew out of the census conflict was momentary and did not spread to other policy areas. Therefore, the spirit of data protection was underdeveloped in comparison to, for example, Germany even after the introduction of the Data Act.

Video surveillance constitutes an exception to this rule. It was unregulated in Sweden until 1977, when the Act on TV Surveillance (Lag om

TV-överakning, SFS 1977:20) was introduced. The Act was a result of the parliamentary commission's report *Photography and Integrity* (Statens offentliga utredningar 1974:85), in which the need for regulation in this area was first established (Prop. 1975/76:194). This was essentially the same committee that drafted the Data Act, and thus there remained the influence of the dystopian imaginaries of inhumane computerisation of the census conflict in the new regulation of video surveillance of the 1970s. The new act implemented a similar system of permissions as in the Data Act, and gave the County Administrative Boards the task of administering both permissions and control of the applications. The basic principle of the act was proportionality between the individual's interest to not be observed and the legitimate interest of control on behalf of the observer; every video surveillance camera had to be approved regardless of who filed the application (i.e., there was no exception for the police, for example; Prop. 1975/76:194, 31). Before a permission was granted, the entity or person who wanted to install the camera had to file an application including the following: reasons for the installation, the technical specification of the camera, the operating hours, aims of the installation, approval of the trade union (in the case of workplace surveillance), and the approval of the local municipality. Apparently, the law was rather restrictive although the principal aim was that video surveillance would generally be approved unless there were good reasons to oppose it—that is, the interests of the general public or, if applicable, employees. The regulation also aimed at ensuring the secure management of installations and footage, since this was the principal threat to citizens (Prop. 1975/76:194, 13). However, the act was also meant to introduce a degree of control and rigidity in the application of video surveillance in order to stop a large-scale deployment, which was considered as a general threat to personal integrity. The system of licensing applied not only to computer-based registers but also to video surveillance, although the Data Inspection Board had no role in overseeing video surveillance practices. For this reason, the institutional practices of video surveillance legislation and personal data regulation developed in quite different ways (see Loftsson, this volume).

West Germany: 'Rasterfahndung' and 'gläserne Bürger'

West Germany's history of surveillance and security was primarily influenced by its experiences of the Nazi regime (1933–45), with its terrible crimes against humanity and the failure of all forms of social solidarity. The West German state was founded on the premise *Nie wieder* (never again), which is reflected in the German quasi-constitutional Grundgesetz (basic law), whose first chapter defines far-reaching civil rights for German citizens and renders the abolishment of democracy constitutionally impossible. It is also reflected in the concept of *streitbare Demokratie* (militant democracy), which denotes the intolerance of the German regime towards undemocratic practices. Created under the auspices of France, the United Kingdom and the United States, West Germany was from the beginning a society imbued with a very strong liberal sentiment. The *freiheitliche demokratische Grundordnung* (free democratic basic order) guarantees citizens very high standards of civil liberties. Against this background it is not hard to understand why German citizens still maintain the most diligent resistance to surveillance.

Similar to Sweden during the 1970s, the Federal Republic of Germany (FRG) was increasingly confronted with fears about the potential dangers of modern information technology, under the heading of 'computerisation'. This was part of a larger international debate on the regulation of personal data and, in 1977, the FRG Federal Data Protection Act (Bundesdatenschutzgesetz, BDSG) was created. In contrast to the Swedish licensing model, the German regulation was based on an advisory function of the Data Protection Commissioner. To a great extent, the BDSG relies on self-enforcement by the registry keeper, especially in the private sector (Bennett 1992: 179ff; Flaherty 1989: 25ff). In the public sector, authorities that process data must report this to the state data protection commissioners, who initiate a negotiation about the data protection measures. For the purposes of this study, however, the German data protection legislation becomes relevant only after the crises of the *deutscher Herbst* (German fall) and the connected population census ruling of the federal Constitutional Court in Karlsruhe in 1983.

From the late 1970s onwards, the German state took a turn towards more repressive policing under the heading of *innere Sicherheit* (internal security), which drew a clear line between the moral establishment and the suspicious activists of the new social movements (greens, peace activists, etc.) that formed in Germany at the time (Hannah 2010). These social movements originated in the student movement and revolts of 1968, which were largely directed against what was seen as the remnants of the fascist state. This was also the backdrop to the terrorist activities of the RAF, which was active from 1970 to 1998, with its peak in the late 1970s and early '80s. During the fall of 1977, the RAF initiated a series of terror acts under the heading 'Offensive 77', which led to one of the greatest crises in FRG history—the *deutscher Herbst* (the German fall).[3] Populist media, in particular the paper *Bild*, built up what has been described as a 'climate of hysteria' in the late 1970s, especially in 1977. Trying to seize the RAF, law enforcement authorities designed new aggressive methods to find the perpetrators, amongst which surveillance was a key feature. One result was an innovative practice called *Rasterfahndung* (dragnet investigation; see Bunderverfassungsgericht 2006a; Kett-Straub 2006). Similar to Clarke's concept of "dataveillance" (1988), Rasterfahndung signified the linkage of different public registers in order to find suspects based on profiles. Activists on the left were subjected to intensive surveillance throughout the 1970s and '80s, as it was generally assumed that many sympathised with the RAF. The boycott movement of the 1983 and 1987 population censuses was subjected to similarly intensive surveillance and repression. This included video and communications surveillance, mail and phone restrictions, illegal confiscations, denial of access to public space, arrests, and the like. Hannah describes these practices and cynically observes that "the zeal, inventiveness, and staying power of state authorities was impressive" (2010: 189), displaying his own astonishment of the severity of the repression. In fact, Kunz (2005) shows us that the politics of surveillance during the entire decade of the 1980s in West Germany was defined by the antagonism between law enforcement and the new social movements. The concept of *innere Sicherheit*, first formulated at the annual conference of ministers of the interior (Innenministerkonferenz, IMK) in 1972, had become a politically loaded term—it signalled both the operational form of *streitbare Demokratie* and an expression of the deep political polarisation in West German society (Kunz cites Jaschke 1991, "Streitbare Demokratie und Innere Sicherheit"; see Kunz 2005: 15, 39).

The discussion above serves to bring us to a specific point in German political history, in which the conflict between the establishment and the extra-parliamentarian opposition reached boiling point. The student revolt, the RAF and the Rasterfahndung method were all involved in the key moment for German regulation of surveillance that still defines the position of the state in relation to the citizenry—the population census of 1983.

Similar to that of the Swedish case (see below), the population census was perceived by the West Germans as a potentially dangerous surveillance measure, giving the state too much power in terms of information. An appeal was made to the Constitutional Court in Karlsruhe, which on December 15, 1983, judged the census to be unconstitutional (Bunderverfassungsgericht 1983). Not only did the court attribute meaning to the census as a means of creating 'glass people' (*gläserne Menschen*), which became a prominent concept in German politics; it also set the standard of 'informational self-determination' (*informationelle Selbstbestimmung*). Embodied in these two concepts—the glass citizen and informational self-determination—the court's decision was based on the constitutional right to free development of the personality, and the argument was that if citizens are under surveillance they will feel too intimidated to fulfil their political rights. The 1983 ruling makes strong assumptions that transcend the traditional legal discourse and borrows ideas and concepts from social psychology and social theory in order to lay out the court's thoughts about the effects of surveillance. Thus both German and Swedish societies shared the fundamental historical moment of problematisation in common: the state's intentions to gain knowledge about citizens, and its practices of surveillance, led to the key moment of contestation and politicisation of surveillance issues and to the regulations that (in part) remain in force today.

It cannot be stressed enough that the ruling restructured German security practices fundamentally—after 1983, all German citizens had the fundamental right to their own data, and surveillance activities that were previously legal now operated in a grey area. All monitoring activities had to acquire legal legitimacy in an ad hoc manner, such as law enforcement practices. The idea of a 'glass citizen' that loses her creativity when under surveillance is a constant theme in the German resistance to surveillance. Informational self-determination also gives the individual citizen the right not to her data as such but to the control over its collection and dissemination (Kammerer 2008: 68). This means that video surveillance became largely illegal because of the difficulties with regard to obtaining the consent of large crowds in open-street surveillance. The census ruling thus became the structuring principle of the FRG regulation of surveillance practices and holds to this day.

I now turn to discussing the socialist states. Due to the nature of socialist rule, it is not possible to discuss surveillance as a problematic practice in terms of democracy, and the analysis here shifts somewhat in a more descriptive direction.

The Surveillance Societies

The socialist states included in this study can rightly be called 'surveillance societies'—societies where social control through surveillance was endemic and integrated in all central societal institutions. East Germany (the German Democratic Republic, GDR) is of particular interest to a study of surveillance, because it represents a modern totalitarian state that utilised every means possible to monitor and control its population. The East German case has in addition been almost completely neglected in the international literature on surveillance, and therefore a closer description of its surveillance practices is warranted. Although these societies do not exist in this form any longer, it is nonetheless important to establish an understanding of the political background in all countries.

East Germany

Surveillance permeated East German society because it was thoroughly integrated into the whole idea of the East German state. As Almgren writes, little was known of the vast extent of the GDR's surveillance machinery before researchers got access to the former state archives after 1990 (Almgren 2009: 98ff, 334ff). Control was inherent in the functioning of every aspect of the GDR state, and went well beyond the security service Ministerium für Staatssicherheit (MfS, in derogative form Stasi) and included all levels and areas of public and private life. East Germany was the most repressive socialist state, albeit not the most violent one. It did, however, manage to completely normalise surveillance more than any other socialist country of its time.[4] Because all state agencies were tasked with guaranteeing total ideological conformity, the Stasi was seamlessly connected with most authorities, and they shared information on a regular basis. For example, the video surveillance of Alexanderplatz and other parts of Berlin ostensibly served as traffic surveillance, but was simultaneously used to monitor dissidents and foreign journalists as they moved around the city (Kowalczuk and RBB 1990). The combination of the GDR's ideological fervour and the developing technology made the regime resemble George Orwell's horrific Big Brother more than did the Nazi regime of 1933–45. The GDR has come to represent the very definition of a 'surveillance society' (cf. Lyon 2001: 33).

Poland

The modern history of Polish politics is a very interesting and turbulent story and, just as in the East German case, it is a history of repression and surveillance, but also resistance. In contrast to the German case, one of the defining characteristics of Poland during socialism was the high degree of political, systemic contestation as early as 1956, and even more importantly in 1980, when Solidarity gathered ten million members—more than a quarter of the entire population. The socialist regime in Poland had suffered from a desperate lack of legitimacy after the demonstrations and riots in 1970, again in 1976, and in particular after the Gdansk demonstrations in 1980–81, which gathered 700,000 workers in protest. The Polish United Workers' Party (Polska Zjednoczona Partia Robotnicza, PZPR) first attempted to cover this legitimacy deficit with economic reforms in the 1970s, but after the Gdansk events there followed over eighteen months of martial law following a decree of the Military Council for National Salvation, instituted by General Wojciech Jaruzelski. This was unconstitutional even by the standards effective at the time, and was seen as a de facto military coup d'état. During martial law and throughout the 1980s, Polish civil society—especially Solidarity—was heavily repressed and several thousand members were imprisoned. Holmes uses the example of Polish martial law as a typical legitimacy crisis during which the regime tries to stay in power by a "reverse to coercion" (Holmes 1997: 53; see also: Avery 1988: 77; Mason 1985: 5ff; Schimmelfenning 1996: 77; Sztompka 2008: 51).

The civil unrest and demonstrations of the 1970s and'80s were all instances of the high potential for political mobilisation among the Polish population. Korek discusses the resistance in terms of "Solidarity discourses" that became a regular and common part of the communication landscape during the 1980s (2004: 47). They had a large influence at the time and still exert their influence in Polish politics, though now more in terms of an *imaginary* rather than *active* movement. In these discourses, a strict distinction between the rulers—*them*—and the people—*us*—was made—a construction that preserved a Polish identity separate from the socialist regime (Wierzbicka 1990: 2). This identity was reflected in semantic practices, and especially in the typically Polish way of expressing both a respectful fear *and* mockery in a single word. Poles feared and loathed the regime but simultaneously ridiculed it, which is evident in terms like *komuna* and *komuch*, which signify the "big, clumsy, shapeless, and rather impotent monster" of the 'real socialism' behind the rhetoric of *komunizm* (the ideology; Wierzbicka 1990: 8). This semantic resistance was used frequently during the 1980s. Below, I briefly discuss the 'mechanics' of the socialist regime's surveillance practices, which reveals the heritage of the Polish and East German populations and makes the current politics of surveillance more understandable.

The Mechanics of the Socialist Regimes

Both socialist regimes were maintained by elaborate systems of monitoring and infiltration. The Stasi effectively governed through an imaginary of fear. Flam has described the resulting personality characteristic as a "fear habitus", by which she means the fear of the Stasi felt by the citizens, whether consciously or subconsciously; fear 'inscribed' on the body and in the 'doubletalk' (the code switching that can be observed in all totalitarian police-states; Flam 1998: 89, 239). Łoś speak about a "taboo mentality" (Łoś 2002: 170). *Infiltration* as a modus operandi is a panoptic technology of power that generates an uncertainty about who can be trusted and about whether or not one is under surveillance (Foucault 1977). The infiltration as a tactic of government in Poland developed less coherently than in East Germany, but the security machinery was nevertheless vast and very real to the Polish population. The Polish security services were known to be more violent and cruel than their counterparts in East Germany, but in total the punitiveness of the Polish regime was less than that of the German one, partly because the Polish Socialist Party was more heterogeneous and irrational than in other socialist countries (Flam 1998: 54f). The infiltration practices meant that fear was generalised and integrated into everyday life. Even though fear might not have been conscious at all times, these are imaginaries that stabilise and solidify political power as they inhibit certain types of behaviour.

The secret services were also tasked with audio and video surveillance of demonstrations, dissident organisations, and individuals in general. The surveillance comprised a range of practices that functioned to maintain the fear habitus. The taboo mentality that Łoś discusses is very similar to the antithesis of the German citizens' 'creativity', and the autonomy mentioned in the context of the 1983 census ruling. Even though there are accounts that claim that this fear habitus eroded in the 1980s (Wierzbicka 1990: 54), I understand these practices of infiltration and surveillance as the main mechanisms that affected social trust, and which therefore are relevant when discussing various forms of surveillance today.

However, beyond certain core aspects, the socialist regimes were quite dissimilar, and here I address some key similarities and differences between them. First, in both countries, the levels of crime were relatively low due to the nature of the police state. However, as Krajewski notes, the crime statistics of the socialist states were notoriously unreliable and probably understated the number of offences (Krajewski 2004: 383, 391). In addition, many deviant activities that would be classified as crimes today were everyday activities, such as certain forms of corruption. Second, the black market was indispensable in sustaining everyday life in Poland because of an economy in which many lived in poverty, especially in the 1970s. In order to survive it was necessary to navigate the highly bureaucratic, socialist system. One of the ways in which this was done was by keeping a healthy dose of economic transfers outside of public attention (Flam 1998: 26ff), which was not as prevalent in East Germany as it was in Poland. Third, the socialist regimes relied on different foundations of legitimacy; whereas in the Polish case the socialist state was not of Polish origin but had been forced onto the people by the Soviets after World War II, the GDR regime enjoyed a high degree of acceptance from the population. Large portions of the Polish population never fully accepted its own regime, which was a reason for its instability and also why the rules had to be comparatively relaxed in certain areas.[5] In East Germany there was for a long time a (fragile) consensus on the general principles of socialism, which was maintained by a very progressive welfare system. Fourth, East German society was highly stable for a long time, in contrast to Poland, which had already by the 1950s experienced social unrest. Whereas the Polish regime became weaker and weaker after each instance of civil unrest (in 1956, 1970, and 1981), East Germany was a fairly stable society until the 1980s. One reason for this was legitimation through semantic manipulation: in what Almgren calls *lingua securitatis*, the discursive practices of the party were centred on a range of antagonistic differentiations, such as 'worker' versus 'bourgeois', 'socialism' versus 'capitalism', 'peace-loving' versus 'aggressive', and 'healthy' versus 'rotting' (Almgren 2009: 453ff). These ideologically antagonistic practices provided easy access to the whole ideological concept of the state, much in the same way as the Bush administration's 'war on terror' antagonised Islam after 2001. They also defined security imaginaries, where the people and the party were united against a common enemy found in the 'rotting' capitalist societies. The Polish socialist regime, on the contrary, was early on under pressure and thus had to introduce economic liberalisations as early as 1956. The GDR was considered to be the 'best student' of the Soviet satellites—even to an extent that the Soviets themselves could not maintain. A common Russian joke at the time was that the GDR was like a chili—the smallest, the reddest, and the fiercest (Wolle 1999: 89). This picture is incomplete, however, and only provides some pieces of a larger puzzle. East Germany was a country of many realities, where legitimation through semantic practices and an extensive welfare system that guaranteed a good life for many citizens was paralleled by intolerance, fear, and repression. Contrary to Pfaff (2001), for example, Almgren shows that good secret informants—and certainly the most important ones—were convinced of the righteousness of their activities. The GDR has been called a 'participatory dictatorship' in the academic literature, which highlights a quite different culture than the Polish attitude of antagonism (Almgren 2009: 32). Many welfare policies of the GDR were quite advanced, such as the extensive parental leave scheme, and the stability of the GDR regime has typically been explained by the relative prosperity and decency of life in what many considered to be one of the most progressive societies at the time.[6] In combination with the surface benevolence of the socialist state that provided certainty, security, and (at least officially) safety, the GDR regime was built on an idea of *progressive humanism*. However incomplete this may have been, the main point is that the two societies rested on very different modes of legitimacy.

This has resulted in some key differences between the now democratic states, particularly in the Polish case (since East Germany was absorbed by West Germany, and was therefore incorporated into an already existing socio-legal system). The radical separation between the people and the state continues to a large extent today, which may be a major explanation for the low levels of trust that Szrubka discusses (this volume). In summary, technical and human means of infiltration and surveillance characterised both socialist states, but to different degrees. Whereas the East German state was very effective and thorough in its control apparatus, the Polish state was brutal but irrational. How this affected these societies after 1989–90 will be the topic of the next section. Immediately below, however, I briefly compare the four country descriptions.

Summary: Regimes in the 1970s and '80s

In the previous section I discussed surveillance and politics in four cases: East and West Germany, Poland, and Sweden. I noted how surveillance and infiltration were key to the functioning of the socialist regimes, but that these rested on different modes of legitimacy. In Sweden, the 1970 population census, along with secret service scandals such as the IB affair, shook society and created fertile ground for a comprehensive regulation of data protection and video surveillance. In West Germany, the antagonism between the new social movements and the extreme Left on the one hand, and the police and conservative political establishment on the other, led to the controversial census ruling of 1983, which was a milestone in German politics in the context of surveillance. Surveillance in West Germany—and Sweden, for that matter—appears as problematic and contradictory: it embodies undemocratic principles both methodically and partly in terms of extent, but at the same time the data protection legislation seems to follow a democratic path. Thus elements of authoritarian, or at least undemocratic, practices can be found in all four societies during this period, whereas a democratically sensitive data protection only existed in Sweden and West Germany. Now I proceed to a discussion on the current situation in Germany, Poland, and Sweden; I will focus on the role of surveillance in general, and video surveillance in particular, and the legacy from the past that informs it.

THE PRESENT-DAY POLITICS OF SURVEILLANCE

The years 1989–90 saw radical changes in Europe, and indeed in the whole world. The fall of the Iron Curtain meant the start of the transformation of Central and Eastern Europe, and new modalities for politicising surveillance in the former socialist countries. The emergence of the European Union (EU) as a crucial regulatory leveller of the field of data protection, and the changed security landscape after 9/11 are two further factors that very much influence the context in which *national* surveillance politics can take place. All three countries have adopted the EU data protection directive EC 95/46, but video surveillance is still a national policy issue. While the legal aspects of this issue are discussed by Jonason in this volume, I want to address here the historical contingency of the current politics of surveillance in light of the previous section. First, I discuss the national cases in the same order as in the previous section; then I compare the cases, and finally discuss the results.

Sweden: Contradictory Movements

The Swedish case is very interesting because of the changes in surveillance regulation that have occurred during the last two decades, and whose contradictory nature cannot easily be missed. During this period, Swedish data protection legislation has been adapted to EU standards twice (in 1998 and 2007), the video surveillance regulations have been gradually liberalised, the Swedish quasi-constitution has been enhanced with privacy statements, and several highly intrusive anti-terror acts have been introduced.

Beyond the strengthening of the—still comparatively weak—constitutional protection of privacy (see Björklund, this volume), the governments in the new millennium have introduced several intrusive acts that enhance law enforcement capacities in the area of surveillance. Examples are electronic monitoring, wiretapping, covert audio surveillance, and data retention. Needless to say, it cannot all be addresses here, but I shall discuss some of these issues in brief below.

The most prominent case was the so-called FRA Act. The National Defence Radio Establishment (Försvarets Radioanstalt, FRA) is a remnant of the World War II and the Cold War, when Sweden had to create signal-decoding techniques. Until recently, the FRA carried out air signal surveillance for a range of Swedish law enforcement agencies as well as the military, but in 2007 the Swedish conservative government presented a bill that would allow the FRA to monitor the Internet as well. Fear arose that this would mean large-scale, generalised surveillance of the Swedish population, and the bill caused a dramatic surge of protest similar to that of the 1970s (Privacy International and Svenonius 2011: sect. 2). The result was a general politicisation of privacy in Swedish politics over the following years, especially in the context of the Internet, about which considerable activism exists in Sweden. Similar imaginaries circulate as those that were portrayed about the state in the early 1970s, although today these are amended to fit the current level of technical sophistication. Yet the same fear of inhumane control and unsolicited surveillance that were discussed forty years ago reappeared during and after the FRA debate. It is, however, an interesting fact that the general politicisation of privacy and the Internet does not extend to the issue of video surveillance (Svenonius 2011, 263). It appears as if the technology remains largely unproblematised in Swedish society, and that the licensing regime guiding surveillance practices is largely automated (see Loftsson, this volume). Licensing has been at the centre of Swedish data protection regulation since the 1970s and its diminishing importance is therefore significant. I therefore discuss this issue further below.

As described above, the window of opportunity for data protection legislation that appeared in the context of the 1970 Swedish census resulted in the Data Act, which prescribed that all registers need permission from the DI. The practice of registry-keeping in Sweden was governed by a notion of consensus, based on regulation through licenses. The DI quickly took on a communicative role in order not to antagonise the registry-keeping agencies and companies it was supposed to supervise because of the weak possibilities for sanction (cf. Bennett 1992: 169, 182). Although the act was increasingly described as inadequate to protect personal integrity (especially in the 1990s), the act and its supervising agency initially seemed to have worked in accordance with the government's intentions, as argued by Markgren (1984: 148, 194).

During the 1990s, the act came under severe criticism, as some argued that the regulation was untimely. During the latter half of the 1990s, as the EU data protection directive was being implemented, a fierce debate raged over the status of data protection. Interestingly, the EU regulation was pitted against the Swedish freedom of information regulation, where critics saw the freedom to post information on the Internet as being threatened by the directive (Olsson 2008). The directive was introduced in Sweden as the Personal Data Act, which kept the licensing regime of the Data Act intact but was so conflict-ridden that a large number of amendments had to be made, and finally the conservative government abandoned the licensing scheme in favour of the internationally more common abuse model in 2007. Ilshammar (2002: 196) observes that the introduction of the EU data protection framework weakened the Swedish data protection regime in favour of a European regime, adapted to the EU internal market. In effect, data protection has been dramatically weakened over the previous twenty years, not only because of the changes to the regulations, but also by the diminishing importance given to privacy enforcement (Loftsson, this volume). Thus the licensing regime of data protection faded away after 2007. However, the licensing regime still thrives in the area of video surveillance. As I described above, the first regulation of video surveillance, Lagen om TV-övervaking, also applied the licensing system. While the data protection regulations were changed in the first two decades of the new millennium, the video surveillance regulations—while also being changed—kept the licenses and so these practices are still relevant for our purposes in this chapter.

The Swedish regulation of video surveillance has also been gradually liberalised during the last twenty years. From the rather strict regulations of the Act on TV Surveillance (1977) and the Act on Surveillance Cameras (1990, which simply introduced some modifications to the 1977 act), via the more generous Act on General Camera Surveillance (1998), to the proposed Camera Surveillance Act, the principle has always been to make surveillance possible while still defending 'personal integrity'. Interestingly, neither this concept nor the threat posed by video surveillance has ever been properly defined in any of the bills (see Björklund, this volume). This makes the judgement about the permissions either one of (a) data protection—that is, the security of storage of video recordings, or (b) the credible possibility that *any* kind of delinquent activity in the surveilled area could override the personal integrity interest. The legal decision by the county administrative boards is therefore a fairly subjective act with a large degree of discretion. As often described in the surveillance studies and criminology, the immaterial value (personal integrity) is usually subsumed by the material value (crime prevention; see Zedner 2003, 2009). Even though the regulation is quite comprehensive, and in some aspects thorough, it has nevertheless made a widespread deployment possible in public areas in Sweden, such as city squares, public transport facilities, and the retail sector. Our research in the West-East Video Surveillance Project also suggests that the Swedish regulations make vast applications of video surveillance possible, and not only in open streets (see Svenonius 2011, 2012). Today there are very few public space video surveillance systems in Sweden—currently, seven systems (Statens offentliga utredningar 2009: 306ff). However, the number of private or semi-public systems such as those on public transport has risen steadily since the 1990s after the regulation of video surveillance was liberalised. Today there are about 18,300 registered systems (Statens offentliga utredningar 2009: 294ff). Whereas we have seen a shift in the Swedish data protection regulations, the video surveillance regulations are characterised by a large degree of continuity. The basic licensing system of the 1973 Data Act is still a key characteristic of today's regulation. Thus regulation of video surveillance can be directly linked to the sensitivity to state surveillance that existed around 1970, and in fact does so much more than the actual data protection legislation that was developed thereafter.

The licensing system does not exist in any of the other countries in this study, and therefore it is a very important aspect of the Swedish case. I now turn to Germany.

'Innere Sicherheit' in Unified Germany

The first two decades of unified Germany were decades of contrasting change in the German security landscape: while quite stable for West Germans, there was a radical change from an East German perspective. After German reunification, there was an intensive debate on how to handle the legacy of communism. The Germans had experienced the problems of 'de-Nazification', but that was a long time before reunification, and the degree of ignorance of conditions in East Germany was high in the former West Germany. *All* Germans experienced Nazi oppression, but since the GDR affected 'only' about seventeen million people it did not have the same gravity in the eyes of many West Germans. This, and the poor economic conditions in the former GDR, have been a constant source of disenchantment for former East German citizens. Further, the 1983 census ruling remains the standard for the German state-citizen relationship, and as I discuss further below, it led the German police to circumvent the standard of informational self-determination by defining 'criminal hot spots' (see Töpfer, this volume). Rasterfahndung has been the subject of several other rulings of the Constitutional Court, including those which occurred in the aftermath of 9/11 as the German federal police were desperate to identify German perpetrators. During that time, all Muslim men with university degrees of a certain age were monitored through Rasterfahndung—something that was later ruled to be unconstitutional by the court in Karlsruhe (Bunderverfassungsgericht 2006b; DeSimone 2010; Lepsius 2004). In sum, Germany saw fewer changes in defence of privacy than in Sweden, but equal effort has been made to increase police surveillance competences, especially during the first decade of the new millennium (Lepsius 2004).

In the area of video surveillance, German social scientists have identified a shift in policing strategy from 1996 onwards, when the first trial commenced in Leipzig of police video surveillance in public spaces. This was closely followed by trials in Westerland and eventually about twenty-five to thirty other cities. The 1996 Leipzig project marked a shift in the way security was perceived (Töpfer 2005). It also marked a politicisation of the individual subject's beliefs and emotions, because a prime focus for police activity became the *feeling* of security, *das Sicherheitsempfinden*. The centrality of video surveillance for *innere Sicherheit* was also emphasised at conferences of the IMK on two occasions, in 2000 and 2006. In 2000, the IMK stated that video surveillance was a suitable means (*geeignetes Mittel*) for crime prevention and crime fighting (Innenministerkonferenz 2000: 27), and in 2006, the ministers urged the German states to increase the use of video surveillance: "The IMK further notes that the instrument of video surveillance is an important contribution to successful counter-terrorism.

With the help of video surveillance suspects can be reliably identified. The IMK expresses its support for an increased usage of video surveillance" (Innenministerkonferenz 2006: 4). This statement was made after a bomb had been found on a train on its way to Koblenz on July 31, 2006. The bomb never exploded due to a malfunction, but the terrorist could be identified with video surveillance footage from the Cologne central station. This event solidified the view of video surveillance as both a crime-preventive and counter-terrorist measure in Germany, particularly on public transport. However, the German citizenry still enjoys the right to informational self-determination, which sets certain limitations to police use of video surveillance in public spaces. In order to be legitimate, the threat to the public's safety must outweigh the need for privacy, and in Germany, unlike Sweden, this has been accomplished through the creation of so-called criminal hot spots (Töpfer 2005). To achieve legality for video surveillance measures, German states (*Länder*) have introduced regulations for high-crime areas and defined these as high-risk zones where the need for security (temporarily) overweighs the need for privacy.

This has, of course, not gone unnoticed. The German autonomous movements that were prevalent in the 1970s and '80s have moved into parliament via the Green Party (Die Grüne), but elements of the autonomous Left were rekindled after the Leipzig video surveillance trials. Contrary to suggestions that German civil society lacked engagement in security issues, the countless seminars, books, demonstrations, citizen initiatives, conferences, flyers, exhibitions, and other forms of protest bear witness to an extensive and active critical movement that state agencies always have to consider in the course of their activities (cf. Chen-Yu 2006). German civil society has the most critical attitude of all the countries in this study, which is historically contingent and reflected in the news media, in civil society, and the production of scientific knowledge. The 1996 Leipzig video surveillance trials were therefore the subject of intense debate, and a range of non-governmental organisations (NGOs) were formed or reinvigorated by the police's surveillance activities. According to Kunz (2005), security policy drew the police's attention away from the Left during the 1990s and early years of the new millennium towards the problems of organised crime, terrorism, and immigration. Kunz's argument is that part of the debate on video surveillance in Germany can be explained by the dislocation of the Left after being 'let alone' in the aftermath of the 1970s and '80s. In that sense, the changes beginning in the 1990s undermined the Left's victim status and new meanings and concepts were created for 'true' underdog status vis-à-vis the 'establishment'; the new 'victims' were the illegal immigrants.

Post-socialist Poland: An Open Field for Surveillance

The most radical break with history among the three countries included in this study was surely witnessed in Poland, whose political system changed abruptly after 1989. While many aspects of these processes are worth mentioning, our prime interest here lies in the unfettered growth of video surveillance that we see today and its historical roots. As Szrubka and Waszkiewicz describe in this volume, video surveillance enjoys very high legitimacy in Poland (see also Svenonius 2011; Waszkiewicz 2011). As Björklund describes (this volume), this high status of video surveillance is related to discourses of modernity and societal progress that are very prevalent in Poland. Both authorities and the general public seem not only to approve but to strongly believe in the preventive effects of video surveillance in situations that would not even be legal in Germany or Sweden, such as monitoring employees. I begin by discussing fear of crime and continue with the regulation of video surveillance—or, rather, the lack thereof.

The 1990s were, for many Polish people, a time of high existential anxiety as the familiar social and political institutions disappeared rapidly and new ones were established slowly and painfully (political scandals became commonplace in the Polish political culture for many years after 1989.) This anxiety soon found its expression in fear of crime (Bauman 2006; Sztompka 2000). Whereas fear was an endemic feature of everyday life during the socialist period, *fear of crime* became equally endemic during the 1990s in Poland (Karstedt 2003; Krajewski 2004; 2009; Łoś 2003, 2005). Karstedt (2003) describes a small surge in murder rates in all transition countries, which began even before the fall of socialism, indicating that the rise in fear of crime may be a reaction to an actual rise in crime rates that were noticeable already before the fall of communism. In the coming years, however, there was a real increase in both 'bad-news media' and in violent crime (Łoś 2002: 166ff). Poland has, in the course of the years after 1989, developed into one of the most punitive countries in Europe, with very high imprisonment rates, high police density, and a vast private security sector (Crawford 2009: 5ff). In the late 1990s, however, a very large disparity between reported fear of crime and recorded crime levels was observed, and towards the end of the 1990s the fear of crime sank while reported crime levels actually rose slightly (Bachmann 2006). At other times, the relationship was reversed. Karstedt (2003) attributes large significance to the police and, in accordance with Łoś and Zybertowicz (2000), the private security sector: "State violence and violent crime as serious violations of human rights have a considerable part in the overall picture of violence in transition states. The police use brutal force in prosecuting crime and criminals. In addition, the privatised security forces, often recruited from former secret police, seem to be quite unrestricted in their use of force against the citizenry." She goes on to state, "The tradition of unrestricted power of the state and its representatives is still prevalent" (305), highlighting how despite modernisation policing remains the same in this respect. The fear of crime is not the same type of fear as during the socialist years because it gains its 'grip' from a different emotional logic, but fear as a social phenomenon and political instrument is a clearly distinguishable feature in both regimes.

What is striking is the anxiety effect of the regime change itself, and how this anxiety was susceptible not only to media reporting, but also to populist rhetoric. Bachmann describes how the politics of law and order, highlighting the high levels of crime and drug abuse, became instrumental for populist parties during the government of Jerzy Buzek that was replaced in the 2001 elections (Bachmann 2006: 282; see also Krajewski 2004). Paradoxically, surveillance became not only the solution to the fear of crime but also a way to progress through modernisation. Surveillance technology, especially video surveillance, has eluded a connection with the repression of the socialist years and has become symbolic of 'progress'—for example, in schools, medical facilities, public transport, and even entire cities. Today, video surveillance is ubiquitous in Polish society. While populist rhetoric aims to reinforce the political logic of *us* and *them*, the *fear imaginary* is hardly related at all to state surveillance. The research mentioned above seems to agree that this imaginary is a kind of manipulation aimed at an entirely different but equally diffuse threat. Just as Bauman (1999) noted, the fear imaginary is a proxy for the uncertainty of transformation. More specifically, however, it is related to modernisation and its sharp distinction between 'winners' and 'losers'.

The issue of video surveillance is seldom a topic of political debate in Poland, and when it is, the issue is whether the costs of expanding surveillance systems can be covered.[7] Also, the regulatory environment is lacking, as there is no proper legislation dealing with video surveillance. In addition, the Polish data protection agency has waited until very recently to intervene in video surveillance. The Polish data protection act has only existed since 1997 and the position of Inspector General for the Protection of Personal Data (Generalny Inspektor Ochrony Danych Osobowych, GIODO) since 1998. In addition, the constitutional protection for privacy is, as Björklund shows in this volume, *formally* much more well-developed than in, for example, Sweden. In a country that is used to a complete lack of transparency in public administration, it is easy to imagine the difficulty in trying to regulate data protection. This is especially true with regard to video surveillance, because of the question of whether video surveillance images constitute personal data or not. The socialist state did have data protection regulations, but not the intention to enforce them. The citizens were not allowed to know what data had been collected about them, or how, why, or what happened to that data; there are indications that some of these practices still remain in both public administration and the private sector. The first inspector general highlighted in particular banks, marketing companies, and the Institute of National Remembrance as problematic areas in Poland (GIODO 2008; Kulesza 2003). Video surveillance, however, has not been viewed as an important area for data protection, and only after Poland's entrance into the EU was the agency pressured to deal with the video surveillance issue (GIODO 2009). The bits of regulation of video surveillance that do exist refer to the personal data act on the one hand and to property legislation on the other. Resonating with earlier mentions of privacy as being closely linked with *human dignity* in Poland, privacy is viewed as personal property. The right to this property, however, is not granted in the public space—a regulation that draws on court rulings from the 1960s and '70s (GIODO 2008). However, the regulation is quite clear in the case of citizens' homes: if a camera films through the window of an apartment this would constitute a serious privacy infringement and the data controller can be prosecuted (Kulesza 2005). The regulation of privacy in terms of property and human dignity thus provides GIODO with some means of regulating video surveillance, but until 2009–10, there had been no discussion on video surveillance regulation. It is only during the last year that the topic has received attention in the public debate (GIODO 2009; Tomczyk 2010). GIODO has yet to engage with video surveillance practices and seems to remain passive to this day, only reacting to notifications of illegitimate systems and notifications from specific video surveillance administrators; such is the strength of the belief in video surveillance in Poland that even data protection officers fail to address the issue.

Summary: Regimes after 1989–90

This second section has described the political situations in unified Germany, Poland, and Sweden from the 1990s to the present. Out of necessity, such descriptions become overviews, and this is true also for this chapter in general. The main topics have been the development of the licensing regime in Sweden, the legacy of the 1983 census ruling for video surveillance in Germany, and the fear of crime in Poland—all of which should be understood in light of their historicity. The licensing regime in Sweden was created to regulate the new computer registers, and was later applied to the regulation of video surveillance, and this proved to be the more stable institution for such an instrument. Whilst the licensing regime was abolished in data protection, it remains in video surveillance. It is interesting to note each country's quite specific way of regulating video surveillance; the Swedish model has been described already. The German model assumes a less powerful role for the state authorities, but on the other hand cases are handled by the more competent German data protection institution. In Sweden the responsible agency is not a specialised data protection institution but an ordinary public office, handling everything from infrastructure to fishing permits. Finally, in Poland, the field is not regulated at all and surveillance cameras are typically installed without consulting a public agency first. The data protection institution, GIODO, has yet to address the issue of video surveillance in a serious way. In a society with such a wide application of video surveillance, and belief invested in it, a thorough regulation as in Germany or Sweden seems far away. In any case, the three countries display distinct approaches to regulation of video surveillance, which arguably stem from the historical factors mentioned in this chapter; this does, of course, not imply that there is no fear of crime in Germany and a complete lack of respect for privacy in Poland, only that the relative importance of these factors varies a great deal.

The suggestion I am making is that during the time frame under examination a number of new factors, some of which were international in scope (such as the end of the Cold War), intertwined with three very different political histories and, accordingly, produced three very different outcomes. These historical backgrounds, as well as the situations today, are key aspects of the research that this volume reports. In the following section I discuss the analyses above from a more general perspective.

DISCUSSION

This chapter takes as its object of interest the historical contingencies of video surveillance regulation and practices. I argue that video surveillance is contingent upon the way that the concepts of *surveillance* and *security* have been framed in each respective society. In Poland, for example, it almost seems as if video surveillance has taken over where the infiltration system left off, despite relatively isolated instances of popular resistance (as Szrubka describes in this volume). The aim of this chapter is thus to analyse the role that events in the past play for current surveillance practices.

Surveillance, particularly video surveillance in public spaces, is a precarious activity for democratic societies because it touches upon such important prerequisites for democracy as privacy, autonomy, and the right of free association (see, e.g., Palm 2007; Ross 1948). Video surveillance is one of the most widespread forms of technologically mediated social control in Western societies. To make disorderly events more visible, it is useful for authorities to be able to turn the clock back and see them again with the (ambivalent) objectivity of a camera image. At the same time, however, video surveillance does not exist in a vacuum but is a technology that is applied to assure stability of the given institutions that use it. This chapter has focused mainly on two types of institutions: data protection and law enforcement. *Data protection* is an idea that is historically and geographically *specific*. It was created as a result of an international debate on computerisation in Western Europe and North America in the 1960s and '70s, but finds little relevance in non-democratic societies with low levels of computerisation in public administration. Data protection, and its primal value 'privacy', thus have a fairly short history and their prevalence today, as analytical and legal concepts, is open for discussion.[8] Nevertheless, because of its strong connection to basic democratic values, the data protection institution is relevant to the purposes of studying the history of video surveillance. The main dynamic here lies in the relationship between people and technology—that is, the extent to which technology is seen as threat that must be contained, or as a way of modernising society. The reason is that regulation is only necessary if the practices that are regulated are seen as problematic. It is a desire for a better future, free from the fear of a brutal, computerised dictatorship that has provided the driving force behind the data protection movement. The functions of data protection agencies and other administrative bodies are reflective of these imaginaries of authoritarianism and technology as an almost utopian vision, and are inherently interesting for any study of surveillance—something neatly displayed by the difference between Germany and Sweden on the one hand, and Poland on the other, in the analysis above.

Law enforcement, in turn, is central to the idea of any social order and is currently the main form of institutionalised and formal social control, even though recent years have witnessed an expansion of security-related practices in many other spheres of society (Garland 2001). In many ways, law enforcement in current Western societies is like data protection: historically and contextually contingent. The police, or similar law enforcement agencies, should (ideally) adhere to the democratic principles as codified in national constitutions. However, since two of the cases included in this study have recent experiences with non-democratic, socialist regimes, these populations have experienced law enforcement and surveillance that adhered to quite different principles. In the socialist regimes, law enforcement in general was subsumed under the more general security apparatus of the Ministry of State Security (in the GDR) and the Ministry of the Interior (in Poland). Accordingly, the cultural sensitivity towards law enforcement might take on different qualities than in Sweden, for example, where the population has had no such experiences. Since the main use for video surveillance is for law enforcement purposes—at least officially—we assumed in this research project that any cultural sensitivity against law enforcement would also extend to forms of surveillance such as cameras. It is surprising how little the socialist past has influenced the issue of surveillance in both Germany and Poland. In both societies, socialism seems to be something that is preferably forgotten—a living memory of a relationship with intrusive technologies. *Forgetting* here means the clear separation between the 'old and outdated' and the 'new modernity' as in the case of Poland, or the assimilation into an existing legal system as in the case of the GDR. Law enforcement seems instrumental in this development, but as the contributions to this volume show, the police are only one among many actors promoting video surveillance. Ultimately, the way in which social control is organised structurally goes beyond law enforcement alone.

In places where video surveillance is traditionally very commonplace—notably, the United Kingdom—the scientific community has started discussing different modes of surveillance, and accounts of more critical attitudes towards video surveillance are starting to engage researchers and gain momentum (see Smith 2012; Fussey 2009). In the three countries included in this study, the extent of surveillance has not yet reached such a point but is heading that way; yet the pace is different: whereas the Polish expansion of video surveillance is extremely fast, and in Sweden moderately fast, the pace in Germany is quite slow (despite the expansion that Töpfer discusses in this volume). Does video surveillance always go through the same phases of expansion, consolidation, and retrenchment? If so, can we expect the birth of a critical debate in Poland? The analyses in this volume are sceptical on this point. Surveillance elsewhere does not seem to have developed in the same way it has in the UK. This should motivate us to conduct further research in a hitherto sparsely researched field, and to perhaps discuss the role of contemporary history a bit further.

NOTES

1. On methods of harassment by East German and Polish police, see Flam 1998: 32ff; on techniques of the Swedish secret service, see Statens offentliga utredningar 2002: 37ff, 213ff.
2. The most notable case was of Leander—a carpenter who was denied employment as a museum technician in the Karlskrona Maritime Museum. Dennis Töllborg, today a professor of law at the University of Gothenburg, defended Leander and wrote several influential texts on the political surveillance activities of the secret services. See Statens offentliga utredningar 2002: 121ff for a detailed description of the Leander case.
3. The actions that took place during the 'deutscher Herbst' included the kidnapping and murder of Hanns Martin Schleyer (former Nazi Party member and subsequently head of the Federation of German Industries), the hijacking of a Lufthansa jet, and the suicides of three RAF imprisoned members; see Schildt and Bundeszentrale für politische Bildung 2001; Spiegel Online 2007.
4. With a *formal* security sector of over 91,000 and an informant network of over 170,'000 so-called *inoffizielle Mitarbeiter* of various classes (1989 figures), the Stasi was the *Schild und Schwert der Partei* (shield and sword of the party; see Gieseke 2000: 54).
5. Flam writes, for example, that the fact that the Polish regime reacted with liberalisations twice (in 1956 and 1970) resulted in the relatively early formation of opposition (1998: xiii, 31ff, 54). The degree to which this was possible varies between different accounts. Whereas some accentuate the repression, others show how, for example, print media and jazz music were ways to mitigate the socialist state's influence.
6. Almgren shows with full clarity how the Swedish establishment endorsed the GDR regime. Most notably for legitimation internally and internationally, Swedish politicians and associations were regular visitors to the GDR. Even conservative newspapers wrote about GDR chairman Walter Ulbricht as a "very sympathetic man," and basic school curricula highlighted the necessity of the Berlin Wall. See Almgren 2009: chap. 1; see also Cooper et al. 2005; and Ommert 2001: chap. 4 on the legitimation through welfare policies in the GDR.
7. I have discussed this elsewhere; see Svenonius 2011: 180f.
8. See the debate in *Surveillance and Society* 8(4), where Bennett (2011b), Regan (2011), Gilliom (2011), Boyd (2011), and Stalder (2011) discuss the notion of privacy based on Bennett's essay "In Defense of Privacy" (2011a).

REFERENCES

Bibliography

Almgren, Birgitta. 2009. *Inte bara Stasi: relationer Sverige-DDR 1949–1990*. Stockholm: Carlsson.

Avery, William P. 1988. "Political Legitimacy and Crisis in Poland." *Political Science Quarterly* 103(1): 111–30.
Bachmann, Klaus. 2006. "Die Liste der vernunft. Polen, der Populismus und die modernisierung wider Willen." *Osteuropa* 11–12: 13–32.
Bauman, Zygmunt. 1999. *In Search of Politics*. Cambridge: Polity Press.
———. 2006. *Liquid Fear*. Cambridge: Polity Press.
Bennett, Colin J. 1992. *Regulating Privacy: Data Protection and Public Policy in Europe and the United States*. Ithaca, N.Y.: Cornell University Press.
———. 2011a. "In Defense of Privacy: The Concept and the Regime." *Surveillance and Society* 8(4): 485–96.
———. 2011b. "In Further Defence of Privacy." *Surveillance and Society* 8(4): 513–16. Boyd, Danah, 2011. "Dear Voyeur, Meet Flâneur . . . Sincerely, Social Media." *Surveillance and Society* 8(4): 505–7.
Bunderverfassungsgericht. 2006a. DFR—Bunderverfassungsgericht 115, 320—Rasterfahndung II.
———. 2006b. Pressemitteilung Nr. 40/2006 vom 23. Mai 2006: Zum Beschluss vom 4. April 2006—1 BvR 518/02—Rasterfahndung nur bei konkreter Gefahr für hochrangige Rechtsgüter zulässig.
Chen-Yu, Lin. 2006. "Öffentliche Videoüberwachung in den USA, Großbritannien und Deutschland. Ein Drei-Länder-Vergleich." PhD diss., Faculty of Social Science, Georg-August-Universität Göttingen. Accessed 6 March 2011 at http://webdoc.sub.gwdg.de/diss/2006/lin/.
Clarke, Roger. 1988. "Information Technology and Dataveillance." *Communications of the ACM* 31(5): 498–512.
Cooper, Hannah L. F., Nancy Krieger, and David Wypij. 2005. "Police Drug Crackdowns and Hospitalisation Rates for Illicit-injection-related Infections in New York City." *International Journal of Drug Policy* 16(3): 150–60.
Crawford, Adam. 2009. "Situating Crime Prevention Policies in Comparative Perspective: Policy Travels, Transfer and Translation." In *Crime Prevention Policies in Comparative Perspective*, ed. Adam Crawford. Devon, England: Willan, 1–37.
DeSimone, Christian. 2010. "Pitting Karlsruhe against Luxembourg? German Data Protection and the Contested Implementation of the EU Data Retention Directive." *German Law Journal* 11(3): 291–318.
Spiegel Online. 2007. "Der Spiegel TV: RAF-Chronik. Der deutsche Herbst." Accessed March 8 2011 at http://www.spiegel.de/video/video-17120.html.
Doyle, Aaron, Randy Lippert, and David Lyon. 2012a. "Introduction." In *Eyes Everywhere: The Global Growth of Camera Surveillance*, ed. Aaron Doyle, Randy Lippert, and David Lyon. Oxon, England: Routledge, 1–19.
———, eds. 2012b. *Eyes Everywhere: The Global Growth of Camera Surveillance*. Oxon, England: Routledge.
Flaherty, David H. 1989. *Protecting Privacy in Surveillance Societies: The Federal Republic of Germany, Sweden, France, Canada, and the United States*. Chapel Hill: University of North Carolina Press.
Flam, Helena. 1998. *Mosaic of Fear: Poland and Germany before 1989*. New York: Columbia University Press.
Foucault, Michel. 1977. *Discipline and punish. The birth of the prison*. New York: Vintage.
Fussey, Pete. 2009. "Beyond Liberty, beyond Security: The Politics of Public Surveillance." *British Politics* 3(1): 120–35.
Garland, David. 2001. *The Culture of Control: Crime and Social Order in Contemporary Society*. Oxford: Oxford University Press.
Gieseke, Jens. 2000. *Die DDR-Staatssicherheit: Schild und Schwert der Partei*. Bonn, Germany: Bundeszentrale für politische Bildung.

Gilliom, John. 2011. "A Response to Bennett's 'In Defence of Privacy.'" *Surveillance and Society* 8(4): 500–504.
Glynos, Jason, and David Howarth. 2007. *Logics of Critical Explanation in Social and Political Theory*. Oxon, England: Routledge.
Hannah, Matthew G. 2010. *Dark Territory in the Information Age: Learning from the West German Census Boycotts of the 1980s*. Farnham, England: Ashgate.
Hempel, Leon, and Eric Töpfer. 2004. *Working Paper No. 15: CCTV in Europe. Final Report*. Urbaneye Project. Berlin: Zentrum Technik und Gesellschaft, Technische Universität Berlin.
Holmes, Leslie. 1997. *Post-communism: An Introduction*. Cambridge: Polity Press.
Ilshammar, Lars. 2002. *Offentlighetens nya rum: teknik och politik i Sverige 1969–1999*. Örebro, Sweden: Universitetsbiblioteket.
Innenministerkonferenz. 2000. "Zur Veröffentlichung freigegebene Beschlüsse der 161. Sitzung der Ständigen Konferenz der Innenminister und -senatoren der Länder am 05. Mai 2000 in Düsseldorf." Accessed 31 March 2011 at http://www.im.nrw.de/inn/doks/imk0500.pdf.
———. 2006. "Beschlussniederschrift über die 181. Sitzung der Ständigen Konferenz der Innenminister und -senatoren der Länder am 4. September 2006 in Berlin."
Joas, Hans, and Wolfgang Knöbl. 2009. *Social Theory: Twenty Introductory Lectures*. Cambridge: Cambridge University Press.
Kammerer, Dietmar. 2008. *Bilder der Überwachung*. Frankfurt am Main: Surhkamp Verlag.
Karstedt, Susanne. 2003. "Legacies of a Culture of Inequality: The Janus Face of Crime in Post-communist Countries." *Crime, Law and Social Change* 40(2–3): 295–320.
Kett-Straub, Gabriele. 2006. "Data Screening of Muslim Sleepers Unconstitutional." *German Law Journal* 7(11): 967–75.
Korek, Janusz. 2004. "Demokrativetande i Polen och Ukraina under den postkommunistiska perioden: En diskursanalytisk studie av nationella, regionala och lokala tidskrifter 1989–2001." PhD diss., Stockholm University.
Kowalczuk, Ilko-Sascha, and RBB. 1990. "Alles unter Kontrolle? Video-Überwachung der Staatsfeinde." *Kontraste—Auf den Spuren einer Diktatur*; accessed 14 March 2011 at http://www.bpb.de/themen/M2NCPV,0,0,Alles_unter_Kontrolle_Video%DCberwachung_der_Staatsfeinde.html.
Krajewski, Krzysztof. 2004. "Crime and Criminal Justice in Poland." *European Journal of Criminology* 1(3): 377–407.
———. 2009. "Punitive Attitudes in Poland—The Development in the Last Years." *European Journal on Criminal Policy and Research* 15(1–2): 103–20.
Kulesza, Ewa. 2001. "Impact of Data Protection Legislation and Activity of the Polish Data Protection Authority on the Practice of Private Sector Entities in Poland." Unpublished manuscript, Generalny Inspektor Ochrony Danych Osobowych.
———. 2003. "The Protection of Personal Data Processes in Public Administration—General Issues." Unpublished manuscript, Generalny Inspektor Ochrony Danych Osobowych.
———. 2005. "Transparency of the State Activity and Data Protection." Unpublished manuscript, Generalny Inspektor Ochrony Danych Osobowych.
Kunz, Thomas. 2005. *Der Sicherheitsdiskurs: die innere Sicherheitspolitik und ihre Kritik*. Bielefeld, Germany: Transcript-Verlag.
Lag (1998:150) om allmän kameraövervakning. Accessed 18 June 2010 at http://www.riksdagen.se/webbnav/index.aspx?nid=3910.
Lepsius, Oliver. 2004. "Liberty, Security, and Terrorism: The Legal Position in Germany." *German Law Journal* 5(5): 435–60.

Łoś, Maria. 2002. "Post-Communist Fear of Crime and the Commercialization of Security." *Theoretical Criminology* 6(2): 165–88.
———. 2003. "Crime in Transition: The Post-communist State, Markets, and Crime." *Crime, Law and Social Change* 40(2–3): 145–69.
———. 2005. "Reshaping the Elites and the Privatization of Security: The Case of Poland." *Journal of Power Institutions in Post-Soviet Societies* 2. Accessed February 6 2012 at http://pipss.revues.org/index351.html.
Łoś, Maria, and Andrzej Zybertowicz. 2000. Privatizing the Police-State: The Case of Poland. Basingstoke: Macmillan.
Lyon, David. 2001. *Surveillance Society: Monitoring Everyday Life.* . Buckingham, England: Open University Press.
Markgren, Sten. 1984. *Datainspektionen och skyddet av den personliga integriteten*. Lund, Sweden: Studentlitteratur.
Mason, David. 1985. *Public Opinion and Political Change in Poland, 1980–1982*. Cambridge: Cambridge University Press.
Merry, Sally Engle. 1986. *Urban Danger: Life in a Neighborhod of Strangers*. Philadelphia: Temple University Press.
Mouffe, Chantal. 1999. "Deliberative Democracy or Agonistic Pluralism?" *Social Research* 66(3): 745–58.
———. 2007. "Artistic Activism and Agonistic Spaces." *Art and Research* 1(2): 1–5.
Norris, Clive. 2012. "There's No Success Like Failure and Failure's No Success at All: Some Critical Reflections on the Global Growth of CCTV Surveillance." In *Eyes Everywhere: The Global Growth of Camera Surveillance*, ed. Aaron Doyle, Randy Lippert, and David Lyon. Oxon, England: Routledge, 23–45.
Olsson, Anders. 2008. *Att stänga det öppna samhället: om ett politiskt missbrukat begrepp: personlig integritet*. Enhörna, Sweden: Tusculum.
Ommert, Horst. 2001. "Kameras helfen nur als Teil einer Gesamtstrategie." *Polizei-heute* 2001: 106–10.
Palm, Elin. 2007. *The Ethics of Workspace Surveillance*. Stockholm: Filosofi och teknikhistoria Kungliga Tekniska högskolan.
Pfaff, Steven. 2001. "The Limits of Coercive Surveillance: Social and Penal Control in the German Democratic Republic." *Punishment & Society* 3(3): 381–407.
Privacy International and Ola Svenonius. 2011. "Sweden—Privacy Profile." Accessed 12 May 2011 at https://www.privacyinternational.org/article/sweden-privacy-profile.
Regan, Priscilla M. 2011. "Response to Bennett: Also in Defense of Privacy." *Surveillance and Society* 8(4): 497–99.
Ross, Alf. 1948. *Varför demokrati?* Stockholm: Tidens Förlag.
Schildt, Axel, and Bundeszentrale für politische Bildung. 2001. "Innere Entwicklung der Bundesrepublik bis 1989." Accessed March 8 2011 at http://www.bpb.de/publikationen/011204358759865033293414441939833.
Schimmelfenning, Frank. 1996. "International Relations and Political Culture: International Debate and Transition to Democracy in Poland." In *The Political Culture of Poland in Transition*, ed. Andrzej W. Jablonski and Gerd Meyer. Wrocław, Poland: Wydawnictwo Uniwersytetu Wrocławskiego, 65–81.
Smith, Gavin. 2012. "What Goes Up, Must Come Down: On the Moribundity of Camera Networks in the UK." In *Eyes Everywhere: The Global Growth of Camera Surveillance*, ed Aaron Doyle, Randy Lippert, and David Lyon. Oxon, New York: Routledge, 46–66.
Stalder, Felix. 2011. "Autonomy beyond Privacy? A Rejoinder to Colin Bennett." *Surveillance and Society* 8(4): 508–12.
Statens offentliga utredningar. 2002. *Politisk övervakning och personalkontroll 1969–2002. Förutsättningarna för säkerhetspolisens politiska registreringar*

och medverkan i personalkontrollen. Report no. 2002:89. Stockholm: Regeringskansliet. Accessed 7 December 2010 at http://www.regeringen.se/content/1/c4/04/56/62852847.pdf.
———. 2007. *Skyddet för den personliga integriteten: kartläggning och analys: delbetänkande.* Report no. 2002:89. Stockholm: Fritze. Accessed 7 December 2010 at http://www.regeringen.se/sb/d/8586/a/79592.
———. 2009. *En ny kameraövervakningslag.* Report no. 2009:87. Stockholm: Regeringskansliet.
Svenonius, Ola. 2011. "Sensitising Urban Transport Security: Surveillance and Policing in Berlin, Stockholm, and Warsaw." PhD. diss., Department of Political Science, Stockholm University.
———. 2012. "The Stockholm Security Project: Plural Policing, Security and Surveillance." *Information Polity* 17: 1–9.
Sztompka, Piotr. 2000. "Cultural Trauma: The Other Face of Social Change." *European Journal of Social Theory* 3(4): 449–66.
———. 2008. *The Ambivalence of Social Change in Post-communist Societies.* Huddinge, Sweden: Södertörn University College.
Tomczyk, Aleksandra. 2010. "Legalność monitoringu wciąż nieuregulowana." Accessed 6 May 2011 at http://www.rp.pl/artykul/212825,511050_Legalnosc_monitoringu_wciaz_nieuregulowana.html.
Töpfer, Eric. 2005. "Die Polizeiliche Videoüberwachung des öffentlichen Raums: Entwicklung und Perspektiven." *Datenschutz Nachrichten* 28(2): 5–9.
Waszkiewicz, Paweł. 2011. *Wielki Brat Rok 2010. Systemy monitoringu wizyjnego—aspekty kryminalistyczne, kryminologiczne i prawne.* Warsaw: Wolters Kluwer Polska.
Wierzbicka, Anna. 1990. "Authoritarian Language in Poland: Some Mechanisms of Linguistic Self-Defense." *Language in Society* 19(1):1–59.
Wolle, Stefan. 1999. *Die heile Welt der Diktatur: Alltag und Herrschaft in der DDR 1971–1989.* Second edition. Bonn: Bundeszentrale für politische Bildung.
Zedner, Lucia. 2003. "Too Much Security?" *International Journal of the Sociology of Law* 31: 155–84.
———. 2009. *Security.* London: Routledge.
Zurawski, Nils. 2007. *Surveillance Studies—Perspektive eines Forschungsfelds.* Opladen, Germany: Verlag Barbara Budrich.

Interviews

Kulesza, Ewa. Data Protection Commissioner of Poland, 1998–2006. Interview by Fredrika Björklund, Elfar Loftsson, Ola Svenonius, Wojciech Szrubka, and Michał Bron. Digital recording. Warsaw, Polish P.E.N. Club. October 3, 2008.
Krasinska, Monika. Director for Jurisprudence, Legislation, and Complaints at the Polish Data Protection Agency. Interview by Fredrika Björklund, Elfar Loftsson, and Wojciech Szrubka. Interview notes. Warsaw, Offices of Generalny Inspektor Ochrony Danych Osobowych. December 18, 2009.

4 The Protection of Privacy in the Context of Video Surveillance
Towards a European Model?

Patricia Jonason

Employed particularly as a means to prevent disturbances of public order in general and serious crime in particular, as well as to gather evidence of crimes, video surveillance allegedly increases public security.[1] However, the use of this technology can also be considered to constitute a threat to citizens' privacy.[2] It is therefore the task of legislatures—both national and international—to balance the interests of security and protecting privacy by putting into place a range of legal safeguards. The issue of video surveillance is highly related to privacy and falls within the scope of human rights protected by the European Convention on Human Rights (ECHR). Sweden, Poland, and Germany are parties to the convention, and thus may not have entirely discretionary power to regulate the installation and deployment of video surveillance. Additionally, because video surveillance constitutes the processing of personal data in the sense of the Data Protection Directive, this European instrument also limits the possibility these three countries—all of which are members of the European Union (EU)—have to decide their own rules concerning video surveillance. However, there is today no supranational, legally binding instrument dealing specifically with video surveillance.

This chapter, which has a legal focus, addresses the issue of video surveillance exclusively from the angle of the protection of privacy. The purpose is twofold. First, we provide an overview of the European legal instruments that deal with the broader right to privacy and data privacy in particular, and we conduct an analysis of the regulation of video surveillance activities in the countries studied in this anthology: Sweden, Germany, and Poland. Second, on the basis of the comparative observations, we discuss whether and in what terms it is possible to talk about a Europeanisation of the legal approach to video surveillance. The concept of Europeanisation is defined here in two ways. The first is as a phenomenon of convergence in which the national legal systems of the European countries become more similar, in part—although not only—due to the influence of European law (including both EU and ECHR law). Second, Europeanisation is understood as the integration of European rules, legal institutions and values into the national legal systems. Needless to say, these two approaches to the concept of Europeanisation are intimately related.

Before turning to the analysis of the legal framework for the protection of privacy, the concepts of privacy and right to privacy must be defined. In addition, the interests at stake as regards privacy in the context of video surveillance must be clarified.

CONCEPTUALISATIONS OF PRIVACY AND OF THE RIGHT TO PRIVACY

The purpose of this section is to briefly present the prevailing European conceptual and legal approach to privacy as well as to clarify the vague and ambiguous notion of privacy. However, the following theoretical attempt to define the concepts of privacy and right to privacy does not pre-judge the fact that neither the European Convention on Human Rights nor the Data Protection Directive—the European legal instruments which constitute the central pillars of the present study—explicitly refer to or contain a proper definition of privacy/the right to privacy.

Different Approaches to Privacy

Privacy is commonly understood to be a panhuman characteristic,[3] and the need for privacy to be a social phenomenon.[4] Still, the level of privacy required by individuals as well as approaches to privacy vary across historical periods and places (Prost 1987: 15). For example, variations between the North American and European conceptual and legal approaches to privacy are somewhat sensitive. In the North American legal system, protection against infringements of privacy mainly focuses on physical privacy in the home and decisional privacy, which includes the right to freedom from government interference in personal decisions (Whitman 2004; Ross 2005). In contrast, in Europe the protection of privacy is more closely associated with the concept of personal dignity (Whitman 2004: 1161 ff). The most eloquent domestic example of the European approach is Germany, where the protection of privacy has been developed within the framework of the larger concept of 'general personality right'. It is intimately associated with the right to dignity, which affords individuals "control over the way they present themselves to the world."[5] However, the existence of common traits in the European approach does not preclude the existence of subtle conceptual and semantic differences among European countries. For instance, in Sweden, the notion of privacy is linked to the concept of 'personal integrity' (*personlig integritet*). This can be compared to the French legal discourse, which conceives of privacy as the 'right to private life' (*droit à la vie privée*), and the German discourse, where privacy is envisaged as a 'right of self-determination' (*Selbstbestimmungsrecht*).[6]

The Concept of Privacy

The concepts of privacy and of right to privacy, which are often merged in the academic literature[7], are not easy to define. Among the plethora of existing definitions and classifications,[8] this study uses Banisar's description, which focuses on the different facets of privacy. According to Banisar (2000), privacy has four facets that are "separate but related":

(1) *Information privacy*, which "involves the establishment of rules governing the collection and handling of personal data"; this concerns data protection.[9]
(2) *Bodily privacy*, which "concerns the protection of people's physical selves against invasive procedures such as genetic tests, drug testing and cavity searches".
(3) *Privacy of communication*, which "covers the security and privacy of mail, telephone, e-mail and other forms of communication".
(4) *Territorial privacy*, which "concerns the setting of limits on intrusion into domestic and other environments such as the workplace or public space". This last aspect of privacy "includes searches, video surveillance and ID checks".

The Facets of Privacy Emphasized in the Context of Video Surveillance

The phenomenon of video surveillance can be split into two elements: the act of *watching/monitoring* someone and the act of *recording* personal information. These elements correspond to the facets of privacy that Banisar labels, respectively, "territorial privacy" and "information privacy", the latter of which is intimately connected to the concept of data privacy.

Recording is more likely to infringe on privacy than surveillance proper, which consists of the observation itself. Indeed, in the context of recording, the captured pictures can be stored, watched, and disclosed eternally and indefinitely. This raises the question of the memory capacity of the data processing technology used for video surveillance. Additionally, recording makes it possible to study the pictures in slow motion or to zoom in; the technology thus permits the watcher/observer to perceive characteristics, details, and the like that the human eye misses in the frame of a 'live' observation *de visu* or on the screen. If, in practice, surveillance/monitoring is often combined with the conservation of the pictures, the distinction between these two elements of video surveillance is important from a legal point of view. This can be seen, for example, in the distinction in German law between *Videobeobachtung* (monitoring) and *Aufzeichnung* (recording; Bornewasser et al. 2008: 226), as well as in the distinction made by the European Court of Human Rights.

The Interests at Stake in the Context of Video Surveillance

There is no doubt that video surveillance, which involves sophisticated technology—for example, it makes it possible to conduct face recognition—raises privacy concerns. In fact, the use of video surveillance technologies might have negative impacts for both individuals and society. As Rouvroy and Poullet (2009: 4) point out, "Under the influence of those "normative technologies",[10] individuals are "decreasingly capable of living by their full autonomous choices and behaviors". Surveillance situations may indeed impair people's ability to develop their personal identities and keep them from engaging in self-improvement. This view was expressed in the German Constitutional Court's census case in 1983, in which some provisions of the Population Census Act were declared unconstitutional. It still permeates the German discourse, as reflected in a decision by the Landtag of Schleswig-Holstein in 2000[11]: "inherent to video surveillance in public space is the risk of social pressure towards conformity that expands far beyond merely acting in accordance with the law, and which obliterates autonomy and freedom". However, video surveillance not only has *immediate* negative impacts on individuals due to its inhibiting effect. In addition, the technical possibilities for retaining personal data that video surveillance technologies make possible might also impact on *the futures* of people subject to surveillance. For instance, there is a plethora of examples of people's lives being damaged because of the publication of their personal data on the Internet.[12]

Beyond the negative impacts that video surveillance can have on individuals, this intrusive technology might also have negative repercussions at the societal level. As the German Constitutional Court acknowledged in the census decision, the fear that information about oneself can be disclosed "would not only impair the individual's chance of development but also impair the common good [*Gemeinwohl*] because self-determination is an elementary functional condition of a free democratic community based on citizens' capacity to act and cooperate"[13].

In any event, despite the existence of different views about the values served by privacy, it is possible to observe "an increasing recognition in academic discourse on both sides of the Atlantic" of the twofold value of privacy (Bygrave 2004: 325; see also Regan 2002)—that is, privacy in its individualistic as well as its democratic form.

THE LEGAL FRAMEWORK FOR THE PROTECTION OF PRIVACY IN THE CONTEXT OF VIDEO SURVEILLANCE

Since video surveillance brings to the fore the need for guarantees for *territorial privacy* as well as *information privacy*—and particularly data privacy—both legal instruments that provide a general right to privacy and instruments dealing with data protection are relevant for the present study. *A fortiori*, instruments specifically regulating the use of video surveillance might be of great importance.

In order to further appreciate the impact that the European legal framework has on national rules protecting the right to privacy and, accordingly, the rules aimed at protecting the privacy of individuals in the context of the deployment of video surveillance, we will examine the two most pertinent regulations at the European level. After that, the national instruments of the three selected countries will be analysed.

European Regulations

The legal mechanisms that protect privacy at the European level and which have the greatest impact on the legal systems of Sweden, Germany, and Poland in matters concerning the right to privacy have been the work of two different organisations.[14] First, they are enshrined in regulations adopted by the Council of Europe, and in particular in the European Convention on Human Rights, adopted in 1950. In addition, they are laid down in legal instruments that have been adopted by the European Community/European Union, especially the Data Protection Directive from 1995.

The European Convention on Human Rights

The legally binding instruments[15] adopted under the auspices of the Council of Europe[16] that are relevant for the present study are actually of two kinds. The first is the European Convention on Human Rights, which guarantees the protection of the *general right to privacy*, and the second is the Convention for the Protection of Individuals with regard to Automatic Processing of Personal Data (Convention No. 108), which deals specifically with *data protection*. The adoption of the latter convention in 1981 responded to the need to take into account technological developments and the requirement that this created for a specific regulation (Krizsán 2001: 59). Ratified by all the member states of the EU,[17] this convention—to which the European Court of Human Rights sometimes refers—has been an important source of inspiration for the adoption of the Data Protection Directive by the European Union. In what follows we will only focus on the European Convention on Human Rights.

The Convention for the Protection of Human Rights and Fundamental Freedoms, commonly known as the European Convention on Human Rights (ECHR), guarantees a number of political and civil rights, including the general right to privacy. The system established by the convention gives citizens the right to submit an application to the European Court of Human Rights (ECtHR) in Strasbourg alleging a breach of convention rights by a contracting state. A violation of the convention principally involves the absence of satisfactory national regulations or the implementation of a national law in a way that fails to respect the rights of individuals. Consequently, decisions of the ECtHR that go against states might prompt them to adopt legislative measures in order to avoid a repeated breach of the convention. They might also influence the practice of the national authorities.

The right to privacy is protected by Article 8 of the ECHR, which states that individuals have the *right to respect for private life*.[18] The interpretation of this concept by the court is extensive and actually encompasses the four facets of privacy as systematised by Banisar. It includes, inter alia, the "right to identity and personal development, and the right to establish and develop relationships with other human beings and the outside world."[19] Nevertheless, as regards issues about private life that arise outside a person's home or private premises, Article 8 of the ECHR does not cover simple monitoring activities. Thus, the court has stated, "The monitoring of the actions of an individual in a public place by the use of photographic equipment which does not record the visual data does not, as such, give rise to an interference with the individual's private life".[20] In order to fall within the scope of Article 8, monitoring must involve "recording of the data and the systematic or permanent nature of the record".[21]

Article 8 of the ECHR offers protection from arbitrary state interference in people's rights (negative obligation), but also obliges the state to protect individuals against interference caused by another private party (positive obligation). Video surveillance emphasizes both kinds of obligations. The state's negative obligation, which requires that it *refrain*—on a vertical level—from acting in a way that infringes on privacy, can be illustrated by the case *Perry v. United Kingdom*.[22] This case concerned covert video surveillance undertaken at a police station in order to make it possible to conduct a video identification line-up at a later date. The ECtHR stated that the interference was not "in accordance with the law" and that Article 8 had been violated. Another case which exemplifies the state's negative obligations is *Peck v. United Kingdom*.[23] Here, the plaintiff had been monitored in the street by a video camera when he attempted to commit suicide. Afterwards, photographs taken from the closed circuit television (CCTV) footage were released to the media. The ECtHR held that the disclosure was not accompanied by sufficient safeguards and therefore constituted a disproportionate and unjustified interference in the plaintiff's private life.

The state's positive obligation, which requires that the state *take positive action* in order to protect privacy within horizontal relationships, can be illustrated by the case *Karin Köpke v. Germany*.[24] In this case, a supermarket cashier was subject to video surveillance by her employer, who suspected that she was manipulating accounts. Taking into consideration the legal and factual circumstances of the case, the court decided that the domestic authorities had struck a fair balance between the plaintiff's right to respect for her private life and the other conflicting interests involved in the case. Therefore, it dismissed the application.

Thus, Article 8 does not recognize an absolute right to respect for private life. Rather, it establishes a conditional right that can be challenged by the existence of other interests, which is typically the situation which occurs in cases involving the use of video surveillance. Interference can therefore be permissible. However, it is only justified if it satisfies the three conditions laid down in Article 8.2. First, the interference must be in accordance with the law. Second, the interference must be in the interest of the legitimate objectives identified in Article 8.2 (i.e., national security, public safety; the economic well-being of the country; prevention of disorder or crime; protection of health or morals; or protection of the rights and freedoms of others). Finally, the interference must be considered to be necessary in a democratic society.

In what follows, using examples related to video surveillance, I will illustrate how the conditions mentioned above have been made more precise by the EctHR.

1. According to the case law of the court, the condition that interference must be "in accordance with the law" comprises two main requirements:
    - A condition concerning the *existence* of a basis in domestic law for the measure. The term law as employed by the ECHR does not signify a law in the formal meaning of the word—that is, that an act emanating from the legislature is required. In the *Köpke* case concerning video surveillance undertaken by the employer at the workplace, for example, the ECtHR acknowledged that the case law of the German Federal Labour Court fulfilled the condition that interference must be in accordance with the law.[25] However, in some cases the ECtHR requires that states adopt legislative provisions in order to comply with their positive obligations under Article 8.[26]
    - A condition related to the *quality* of the law. The court requires that the domestic legal basis must satisfy high standards related to the rule of law. The legal basis must therefore be "accessible"[27] and "foreseeable".[28] In matters regarding secret surveillance the court additionally requires that the law contain guarantees against arbitrary interference. In the *Köpke* case the court found this condition fulfilled, because the case law of the Federal Labour Court, "by interpreting the scope of the employee's fundamental right to privacy as guaranteed by Article 2 § 1, read in conjunction with Article 1, of the Basic Law, developed important limits on the admissibility of such video surveillance, which safeguarded the employee's privacy rights against arbitrary interference."[29] Additionally, in order to be admissible, the impugned measure must conform to the requirements set out in the domestic law which permits the interference.[30] This requirement was not met, for example, in the *Perry* case concerning the identification of a person prosecuted for robberies by means of covert videotaping for use in an identification line-up.
2. The second condition for the admissibility of the interference is that it must be in the interest of the legitimate objectives identified in Article 8.2. Several of the interests listed in Article 8.2 may be invoked to justify permissible interference with an individual's privacy by means of video surveillance. For example, it can be justified on the grounds of the need to prevent disorder or crime as well as to protect national

security or to protect the rights and freedoms of others; property rights were referred to in the *Köpke* case.
3. The requirement that an interference must be a necessity in a democratic society implies that the interference corresponds to a "pressing social need"[31] and is "proportionate to the legitimate aim pursued".[32] The member states—through their national or local authorities—are generally in a better position than the European Court of Justice to decide what kind of measures are necessary to serve the relevant interests. However, their power of appreciation is not unlimited, but subject to the supervision of the court. The application of the principle of proportionality as regards adopted measures—"a balancing approach that requires the intensity of the restriction not to be excessive in relation to the legitimate needs and interests which gave rise to it" (Tsakyrakis 2008) must, inter alia, take into consideration whether less-intrusive measures were available and could have been used to achieve the desired goals (Clayton and Tomlinson 2000: 278). In the above-mentioned *Köpke* case, the European Court—taking into account (1) that the surveillance was limited in time as well as geographically, (2) that the data that was recorded was processed by a limited number of persons, and (3) that the domestic courts had concluded that there were no less intrusive means available to achieve the desired goal—declared that "there is nothing to indicate that the domestic authorities failed to strike a fair balance, within their margin of appreciation, between the applicant's right to respect for her private life under Article 8 and both her employer's interest in the protection of its property rights and the public interest in the proper administration of justice".[33]

However, it's important to note that the case law of the European Court of Human Rights does not exhaustively cover the video surveillance issue. On the contrary, the guidelines set out by the court reflect its fragmented view, which is a result of the fact that the court can only adjudicate the cases submitted to it.

The Documents Adopted by the European Union: The Data Protection Directive

Among the legally binding tools adopted within the framework of the European Community/European Union, Directive 95/46/EC of the European Parliament and of the Council of 24 October 1995 on the protection of individuals with regard to the processing of personal data and on the free movement of such data (the Data Protection Directive) is particularly relevant here, and it is on this instrument that we will focus. It is interesting to mention, however, that the Charter of Fundamental Rights of the European Union contains provisions related to the right to privacy.[34] These are Article 7, which provides for the protection of private and family life, home and communication, and Article 8, which guarantees the protection of personal data.

The purpose of the Data Protection Directive is to harmonize EU member states' legislation related to data protection in order to allow the free flow of personal data within Europe and avoid trade obstacles while protecting individuals' privacy. It contains rules regulating the processing of personal data and is applicable to both the public and the private sectors. In what measure has this directive had an impact on national legislation regarding video surveillance? The question is actually twofold and the answer depends on the field of application of the directive as regards, first, the technology covered and, second, the purposes served by the processing.

1. The Directive—in particular, Paragraph 14 of the preamble, states that it is intended to apply to sound and image data. Interestingly, the directive covers video surveillance. Interestingly, the directive takes into account the specificity and sensitivity of the processing of sound and image data and the technological developments that this kind of processing is subject to.[35]
2. However, as the directive does not apply to "processing operations concerning public security, defence, state security and the activities of the state in areas of criminal law" (Article 3.2), activities that are typical for video surveillance performed by public actors do not fall within the scope of this European instrument. However, the member states have a discretionary right to apply the directive to matters that are explicitly outside its scope. In practice, some national legislatures have allowed it to have a spillover effect on national rules.

By posing conditions for the processing of personal data, the directive aims to prevent infringements on the data subject's private life. The directive lays down *principles relating to data quality* (Article 6). The data must be processed fairly and lawfully, and must be collected for specified, explicit, and legitimate purposes. In addition, it must not be processed further in a way that is incompatible with those purposes (the finality principle). The data may not be kept longer than is necessary for the purposes for which it is processed (the right to be forgotten). Furthermore, the directive sets out *criteria for making data processing legitimate* (Article 7). Either the data subject has given his or her unambiguous consent to the processing of data, or processing is necessary for a goal mentioned in the directive (i.e., contractual obligations, compliance with a legal obligation, the protection of the data subject's vital interests, the performance of a task carried out in the public interest or on the basis of balancing of interests). The processing of so-called sensitive data (e.g., data concerning health or revealing racial or ethnic origin) is prohibited in principle (Article 8). According to the directive, *information* must be given to the data subject. The controller must provide the data subject with certain information, including for instance the identity of the controller, the purposes of the processing and the recipients of the data(Articles 10 and 11). In addition, the Data Protection Directive specifies the rights of data subjects. These include the right of *access to data* (Article 12) and the right to *object* to the processing on compelling legitimate grounds (Article 14a). Furthermore, the directive contains provisions on safeguards that apply in connection with *automated individual decisions* (Article 15), *security* of processing operations (Article 17), *notifications* (Articles 18 and 19) and *prior checking* of processing operations (Article 20),[36] as well as provisions on the *transfer of data to third countries* (Articles 25 ff).

Even if all of these provisions are, *a priori*, applicable to the processing of sound and image data,[37] the fact remains that some provisions pose problems in their concrete application,[38] not least in the context of video surveillance. For instance, access to data raises the question of the right to data protection of third parties, whose images might be included in the same processing. Other questions concern the content of the obligation to inform (Articles 10 and 11) and the obligation to obtain the consent of the data subject (Article 7a). Regarding the latter question, is it possible to consider the subject to have consented when he or she passes by a video camera?

Rules of National Character

This section aims to provide an overview of the Swedish, Polish, and German instruments covering the regulation of open-street surveillance deployed in places where the general public has access. It will outline the similarities and differences among the national regulatory measures. Additionally, the following analysis will serve to place the national rules regarding video surveillance in relation to the European Convention on Human Rights and the Data Protection Directive in order to discern the extent of the phenomenon of Europeanisation in the area of video surveillance regulation.

First, we have two methodological remarks: (1) In order to make the comparison manageable, we will limit its scope to open-street surveillance (in contrast to secret/covert surveillance). Among the rules regulating open surveillance, we will focus on general rules and exclude sectoral ones (e.g., specific rules concerning video surveillance undertaken by police authorities). (2) The following examination will also cover Swedish data protection legislation and not just the Act on Public Video Surveillance (1998:150), although only the latter specifically applies to video surveillance of public spaces. This choice, which, incidentally, facilitates the analysis of the process of Europeanisation of national legal solutions, is justified by the fact that current reform proposals in Sweden in the area of video surveillance explicitly lay down the application of data protection rules to video surveillance used in places to which the general public has access.

The following description and analysis of the Swedish, Polish, and German legal texts regulating video surveillance will particularly emphasize the *guarantees* established to protect against infringements of privacy that may occur because of the use of video cameras. In order to better understand the place and value granted to the right to privacy in each of the three countries, as well as the constitutional safeguards surrounding the right to privacy, the country analysis will begin with a short review of the constitutional provisions germane to this right.

The Swedish Legal Framework

The *constitutional* protection of privacy in Sweden is quite weak in comparison to other countries—not least compared to Poland and Germany. Additionally, in contrast to the ECHR, the Swedish Constitution assigns low value to privacy.[39] Indeed, the Swedish Constitution—that is, the Instrument of Government (Regeringsformen 1974:152)—contains no legally binding rules offering *general* protection for privacy.[40] Thus, the approach adopted by the Swedish legislature is characterized by a fragmented constitutional guarantee of individual privacy against state interference, combined with guidelines requiring the state to act in order to protect privacy/the private life.[41] The former (i.e., the constitutional guarantee) includes protection against violations of privacy in several sensitive situations, including examining mail and other confidential correspondence, eavesdropping and recording telephone conversations, corporal punishment, and torture and medicinal influence aimed at extorting or suppressing statements.

Another factor which contributes to the relative weakness of protection of privacy in the Swedish legal order is the fact that the constitutional provisions were not primary motivated by the value assigned to protecting privacy *per se*, but rather to protect the free formation of opinion.[42] However, influenced by the ECHR, the Swedish legislature has recognized the need to take into account the value of privacy itself, and to reinforce protection for privacy at the constitutional level. This increased awareness resulted, in 2011, in the insertion of a provision about surveillance into the catalogue of rights in Chapter 2 of the Instrument of Government. The new provision (Chapter 2, §6, 2st) states that "everyone shall be protected in their relations with public institutions against significant invasions of their personal privacy if these occur without their consent and involve the surveillance or systematic monitoring of the individual's personal circumstances". The goal of the new provision is to protect information about an individual's personal circumstances. Information in this context is understood to include various types—for example, photographs.[43] The new provision protects individuals against surveillance measures (e.g., the collection of data undertaken by public authorities in order to make administrative decisions[44]), as well as measures consisting of monitoring (e.g., secret video surveillance and activities falling within the scope of the Act on Public Video Surveillance 1998: 150).[45] However, the decision of the legislature to put the provision protecting individuals against surveillance under the heading of "Bodily integrity and freedom of movement" creates some perplexity concerning which fundamental right is supposed to be protected by this new provision.

Another change related to privacy that is significant in relation to the phenomenon of Europeanisation is the Swedish legislature's decision to remove a provision concerning data protection from the constitution. In support of its abolition, the legislature invoked, inter alia, the fact that, not least due to the Data Protection Directive, Sweden has a legal obligation to ensure data protection regulation even if it is not explicitly expressed in the Constitution.[46]

As we can see, the influence of the European legal framework on the Swedish legal system in the matter of privacy protection is unquestionably tangible. However, the Europeanisation process manifests itself in opposite ways depending on whether we are talking about the impact of rules stemming from the Council of Europe (here the ECHR in particular) or the influence of rules emanating from the legal framework of the European Union—in particular, the Data Protection Directive. While the increasing influence of the ECHR on the regulation of privacy protection is indicated by the introduction of a new provision in the Swedish Constitution, the impact of EU law resulted in the elimination of a constitutional provision. The former reflects a wish to bring the protection for privacy offered by the Swedish constitutional provisions better in line with the protection provided in Article 8 of the ECHR and its interpretation by the ECtHR.[47] The latter reflects the growing achievement of the integration of law emanating from the European Union into Swedish law. Regardless of the changes to the constitution, the protection of privacy in Sweden has improved since the ECHR was incorporated into the Swedish legal system on 1 January 1995.[48] Since then, citizens may refer to Article 8 of the convention in domestic courts. Moreover, the Swedish legislature is constrained to respect the European Convention on Human Rights, and not least Article 8 on the right to privacy, when enacting new laws (Chapter 2, §19).

The Regulation of Video Surveillance

Open video surveillance activities in Sweden are regulated by two pieces of legislation: the Act on Public Video Surveillance (1998:150) and the Personal Data Act (1998:204). A new piece of legislation concerning video surveillance which is partly a synthesis of these two acts has been proposed.

The Act on Public Video Surveillance was passed in 1998,[49] and is applicable to all kinds of open video surveillance. However, when video surveillance is conducted in places where the public does not have access and in semi-public areas (e.g., schools and residential areas), only a few provisions of the law are applicable. In these situations, the Personal Data Act might apply, since surveillance implies the processing of personal data.

The Act on Public Video Surveillance states in its portal paragraph that video surveillance must observe proper respect for personal integrity (*personlig integritet*), and the legislation contains different guarantee mechanisms in order to fulfil this goal. First, the act requires authorization and notification. Thus, the installation and use of CCTV in places accessible to the general public must generally be authorized in advance by an administrative authority, the County Administrative Board (*länsstyrelsen*; see §5). In some cases, all of which are listed in the act, only a notification to the same authority is required (§§11–12).[50] In some other situations, there are no formal requirements (§§7–10). The act lists the information that permit applications and notifications must include (§§16–17). In the case of requests for authorisation, the information required includes the purpose of the video surveillance, equipment to be used, where the equipment will be installed and the space to be monitored. This helps the County Administrative Board decide whether or not the video surveillance request should be approved, which depends on whether it complies with the condition set out in Section 6, which states, "Authorisations are only given when the interest of surveillance weighs more than the interest of individuals not to be subject to surveillance".

A second kind of guarantee of privacy provided by the Act on Public Video Surveillance consists of the duty incumbent upon the entities that deploy CCTV to mark the area under surveillance with "clear signs or in another effective way" (§3). In cases where sound is recorded, this fact must also be explicitly stated. The obligation to inform about surveillance also applies in cases where authorisation for surveillance is not needed, as well as in areas where the public may not have access. Additional guarantees in support of respect for privacy consist of provisions about who is entitled to have access to the recorded material (§13) as well as about the retention of this material—specifically how long it can be saved—and its destruction (§14).[51]

The second Swedish law of general character which is applicable to non covert video surveillance is the Personal Data Act (Personuppgiftslagen, PuL). It was passed 1998 in order to transpose the European Data Protection Directive into Swedish law.[52] Video recording using digital cameras, regardless of the limitations regarding areas of application which are specified in the directive, has been seen as constituting the processing of personal data as defined in the PuL. Thus, it falls within the scope of the act to the extent that it is not specifically regulated in the Act on Public Video Surveillance.

The protective mechanisms contained in the act and derived from the Data Protection Directive are applicable to the processing of personal data in the form of recording by means of video surveillance. These include principles related to data quality (§9), criteria making the use of data processing legitimate (§10), and the requirement that the data subject must be informed (§§23–27). In any event, the obligation that the controller must inform the Data Inspection Board (Datainspektionen)—the supervising authority—of the processing activity is rarely applicable because of a plethora of exceptions. In addition, the PuL was amended in 2007. The traditional regulatory model (*hanteringsmodell*) consisting of the regulation of every step in the processing of personal data was replaced by an abuse-centred regulatory model (*missbruksmodell*) in situations involving the processing of personal data in *unstructured material*. This shift has had great consequences for the conditions surrounding video surveillance. Indeed, as the audio and video recordings emanating from surveillance cameras have been considered to consitute processing of personal data in *unstructured material*,[53] the majority of the provisions of the PuL—including, inter alia, the above-mentioned provisions—are thus not applicable to this kind of processing. Consequently, the protection mechanisms for privacy[54] set out in the PuL are not, for the most part, applicable to surveillance that takes place in areas in which the public does not have access, including semi-public areas. And although there is a safeguard—that is, the prohibition of the processing of personal data in unstructured material which entails a violation of the personal privacy of the person (§5a)—the application of the abuse model on video surveillance has been viewed as problematic from a privacy perspective.[55]

Proposal for a New Act: The Act on Video Surveillance

In 2009, a parliamentary committee tasked with evaluating the current legal framework on video surveillance made proposals aimed at increasing the level of protection of privacy, as well as paying more attention to the Data Protection Directive in the context of video surveillance activities.[56] A new piece of legislation has been proposed. It is intended to cover all public video surveillance except that which is undertaken for private purposes, and it brings together provisions from the current Act on Public Video Surveillance and some rules from the Personal Data Act.

In order to remedy the inconvenience of the application of the abuse model on video surveillance in *places where the public has no access* and to better take into account the Data Protection Directive,[57] the report contains proposals to neutralize the application of the Personal Data Act and proposes instead the setting out of specific provisions for this kind of processing. The consent from persons under surveillance is required, which presupposes that those who undertake video surveillance must identify themselves, explain the purpose of the video surveillance, and provide information about whether surveillance includes audio and/or visual recording. Without consent, video surveillance is only permissible in order to prevent crimes or accidents or for other legitimate interests in cases in which the benefits of video surveillance outweigh the interests of individuals not to be monitored. Furthermore, the proposal specifies that the purpose of the video surveillance must be specific and explicit and that surveillance must not be more comprehensive than what is required in order to fulfil the goals.

Concerning video surveillance in *places which are accessible to the public*, the mechanism of licensing and notification that exists in the current Act on Public Video Surveillance will be retained. Nevertheless, the legislature wants to reinforce protection of privacy by means of provisions derived from the data protection legislation. Indeed, it has been proposed that the new act contain provisions on the general conditions for undertaking video surveillance. One is the introduction of the finality principle, which states that images and sound data may not be processed for a purpose incompatible with the original purpose. Others are security provisions in the form of technical and organisational measures. These provisions are to be applicable both to video surveillance in places to which the public has access as well as places that are not publicly accessible.

The Polish Legal Framework

The Polish Constitution adopted in 1997 recognizes the right to privacy in several provisions. Article 47 provides a general right of legal protection for private life; Article 49 states that there is a right to privacy of communication; and Article 51 establishes the right of data protection. Additionally, Article 30 proclaims the inviolability of the dignity of the person and Article 50 the inviolability of the home. Especially of interest for the current study is the fact that the Polish Constitution, like some other constitutions adopted in modern times,[58] recognises an autonomous right to data protection. However, the guarantees surrounding data protection are remarkable for their precision, which is uncommon at the constitutional level. The constitution establishes explicitly negative obligations on the state: "Public authorities shall not acquire, collect nor make accessible information on citizens other than that which is necessary in a democratic state ruled by law."[59] It guarantees the individual the right of access to data concerning him- or herself, as well as the right to demand that untrue or incomplete information, as well as information acquired by illegal means, are corrected or deleted.

The provisions guaranteeing the right to privacy have been used in order to constitutionally challenge an act which authorized video surveillance. Indeed, a group of members of the parliament referred the 2002 Internal Security Agency and Foreign Intelligence Agency Act to the Polish Constitutional Tribunal, arguing that provisions related to the functions and activities of the agencies mentioned in the act were incompatible with the constitution. The tribunal proclaimed the provision to be unconstitutional.[60] In the ruling, it noted that the provision empowering officers of the Internal

Security Agency to monitor and record events in public places without the knowledge or consent of those concerned severely interfered with civil rights and freedoms and called attention to several defects—for example, the absence of guarantees for ascertaining that the provision would only be used to realise the goals defined in the act. The tribunal ruled that Article 23.1.6, which empowered officers to observe and register images of public events, violated several constitutional provisions, including Article 2 on the rule of law, Article 30 on dignity, Article 47 on right to legal protection of private life, Article 49 on privacy of communication, and Article 51(2) on personal data protection, read in conjunction with Article 31(3) on limitations of constitutional rights and freedoms.

Unlike in Sweden there is no legislation of general character specifically regulating public video surveillance activities in Poland.[61] However, legislation about data protection applies. The Act of 29 August 1997 on the Protection of Personal Data (APPD), which constitutes the transposition measure for the Data Protection Directive 95/46/EC,[62] does not explicitly regulate the processing of sound and image data or video surveillance. However, the public authority in charge of the application of the APPD, the Polish Inspector General for Personal Data Protection (Generalny Inspektor Ochrony Danych Osobowych, GIODO), acknowledged in a discussion of the issue of recording image and sound by CCTV deployed in the workplace that the scope of the application of Polish law includes video surveillance.[63] Making reference to Directive 95/46/EC, and specifically to the preamble, Paragraph 14, GIODO stated that Article 6 of the APPD, which defines the notion of personal data, covers visual data. Thus, the controller is required to fulfil all the obligations resulting from the provisions of the Polish act.

The guarantees applying to the controllers are similar to the Swedish ones enacted in the PuL, because they also derive from the EU's Data Protection Directive. They take the form of "Principles of Personal Data Processing" (Chapter 3), which contain the criteria that make data processing legitimate. Like the Swedish data protection act, the principles related to data quality are specified in the form of obligations addressed to those who control video surveillance (Article 26). Furthermore, the Polish act obligates the controller to inform all concerned persons (Articles 24 and 25), take security measures and notify the inspector general of the existence of the data filing system (Chapter 5). Making decisions in an individual case based solely on information gathered by automated data processing is prohibited (Article 26a). The Polish act also provides for the rights of data subjects (Chapter 4).

The German Legal Framework

The German Constitution, the Basic Law for the Federal Republic of Germany, was promulgated in 1949. Like its Swedish and Polish counterparts, it contains provisions aimed at protecting certain aspects of individual privacy (Article 10, Privacy of Correspondence, Post, and Telecommunications;

Article 13, Inviolability of the Home). A general right to privacy and non-interference in personal affairs has been build on the basis of Articles 1 and 2, which guarantee, respectively, human dignity and personal freedoms (Young 2006: 161). Conceptually, video surveillance is said to lead to an infringement of the general right of personality in its manifestation as the fundamental right to informational self-determination (*informationelle Selbstbestimmungsrecht*), as guaranteed by Article 2.1 of the Basic Law in conjunction with Article 1.1.[64]

Weichert[65] noted in 2001 that the principles acknowledged in the 1983 census decision, called "the most important decision in the history of German data protection" (Hornung and Schnabel 2009: 85), also apply to video surveillance. These principles are: the existence of a general interest to permit restrictions on the right of informational determination, the existence of a clear legal basis for doing so, and proportionate interference with privacy. A decision by the Constitutional Federal Court on 23 February 2007 dealing specifically with video surveillance confirms these requirements and particularly the need for a clear legal basis.[66] In this judgement, which concerns video surveillance deployed by the City of Regensburg in order to prevent the vandalism of a piece of art displayed on the ruins of a synagogue, the court declared that the decision of the Bavarian Administrative Court, which had confirmed the decision of the municipality to deploy video surveillance, violated the complainant's fundamental rights under Article 2.1 in conjunction with Article 1.1 of the Basic Law.[67] According to the Constitutional Court, this use of video surveillance—consisting of the monitoring of places publicly accessible and video recording of persons, the majority of whom had not committed an act of misconduct—amounted to significant interference with the right to self-determination. The legal basis of the decision—Articles 16 and 17 of the Bavarian Data Protection Act, which contain only general guidelines—was therefore unsatisfactory. On the contrary, a definite and clear legal basis was required.

Following the decision of the Federal Constitutional Court, the Bavarian legislature introduced a specific provision in the Bavarian Data Protection Act, Article 21a, titled "Video Monitoring and Video Recording (Video Surveillance)". Particularly worth noting in the court's judgement is its emphasis on the risks of violations of privacy due to technology, which notes that "through recording, the life [*Lebensförgänge*] of persons subject to monitoring will be technically fixed and as a result can be retrieved, processed and evaluated as well as linked with other data. Thus a variety of information about identifiable persons can be obtained which can, in extreme cases, lead to the profile of behaviour . . . of the persons concerned" (§38).

The Regulation of Video Surveillance

In Germany, public video surveillance is regulated in data protection legislation both by specific rules dedicated to video surveillance and by general data protection rules.⁶⁸ Since Germany is a federal state, both legislation at the federal and länders levels is applicable. The application of the legislation at the different levels is as follows. The Federal Data Protection Act applies to the federal public sector and the private sector, while the data protection acts adopted by the *Länder* (states) apply to their respective public sectors. The Federal Data Protection Act is supervised by the Federal Data Protection Commission (Bundesbeauftragter für den Datenschutz), which is an independent federal agency. At the state level, each *Länder* has a data protection commissioner to supervise the state data protection acts and the private sector.

We will now examine, at both the federal and state levels, the *specific provisions* dedicated to video surveillance of publicly accessible places and the *general data protection rules* contained in the data protection legislation. The legal framework from Schleswig-Holstein has been chosen, in an arbitrary manner, to illustrate the pertinent legal framework of the länder.

The German Federal Data Protection Act (Bundesdatenschutzgesetz, BDSG) was adopted 18 May 2001 in order to make the German legislation consistent with the EU's Data Protection Directive.⁶⁹ It contains a specific section devoted to video surveillance—Section 6b, "Monitoring of Publicly Accessible Areas with Optic-electronic Devices". There are different kinds of protection mechanisms contained in this section in order to prevent infringements on privacy/data protection privacy caused by video surveillance.

A first guarantee consists of limiting the use of video surveillance for certain purposes, which are listed in the law itself. According to Section 6b, surveillance is "allowable only in so far as it is necessary" to:

1. fulfil public tasks
2. exercise the right to determine who shall be allowed or denied access
3. pursue legitimate interests for specifically defined purposes.

Furthermore, Section 6b(2) requires that suitable measures be taken to make it clear that an area is being monitored, as well as to identify who is doing it). In addition, "where a specific person can be identified from data collected through video surveillance, the person is to be notified about the recording or use of it in accordance with Sections 19a and 33". These sections specify notification requirements as regards the subject of data collection and constitute typical data protection mechanisms. Other protection rules are those laid down in Section 6b(3), which states that data "may be processed or used if necessary [i.e., there are no other means available] to achieve the intended purposes and if there are no indications of overriding legitimate interests on the part of the data subject. These data may be processed or used for another purpose only if necessary to prevent threats to state and public security or to prosecute crimes". The same can be said about Section 6b(5). Here it is stated, "The data shall be erased as soon as they are no longer needed to achieve the purpose or if further storage would conflict with legitimate interests of the data subject".

The Data Act of Schleswig-Holstein (Landesdatenschutzgesetz, LDSG) is from 2000 and, like the Federal Data Protection Act, contains specific provisions regarding video surveillance. These are found in Section 20 on "video monitoring and recording". According to Section 20(1), publicly accessible areas may be monitored with optic-electronic devices "in so far as is necessary" to:

1. fulfil public tasks
2. protect one's own property (*Hausrecht*).

However, this is conditioned with the provision "and the legitimate interests of those concerned do not predominate".

According to Section 20(2) the material from monitoring may be stored (e.g., a video recording) if those being recorded are aware of it—that is, if they can discern by appropriate means that they are being recorded. At the latest, the data is to be deleted after seven days unless the video surveillance has documented incidents which must be clarified, thus justifying longer retention.

As we can see, both laws make use of the protection mechanisms that are typical for video surveillance legislation. This includes the introduction of a *specific list of purposes* for which video surveillance may be used. However, the list in the BDSG is more extensive than the one in the LDSG. Even if this can be explained by the fact that the federal law has a wider scope of application because it covers both the public and the private sector, the fact remains that the concept of 'legitimate interests' used in the federal legislation is "unnecessarily far-reaching and flabby".[70]

The general protection rules stemming from the Data Protection Act are a complement to the specific rules concerning video surveillance in publicly accessible places and are applicable only in cases of video surveillance in non-public places. These rules are similar to the ones laid out in the Data Protection Directive. Both the BDSG and LDSG contain rules related to the right of access to data (BDSG §§19 and 34; LDSG §27[71]); rights to rectification, erasure, or blocking (BDSG §§20 and 35; LDSG §28); prohibition against making decisions based solely on automated processing of personal data (BDSG §6a; LDSG §19); lawfulness of data collection, processing, and use (BDSG §4; LDSG §11); and the finality principle (BDSG §3a, Data Reduction and Data Economy; LDSG §13).

Comparison and Reflections on the Phenomenon of Europeanisation

In this section we aim to briefly compare the legal frameworks of the three countries of Sweden, Poland, and Germany before reflecting about the phenomenon of the Europeanisation of the regulation of video surveillance.

A Comparative Approach

The analysis of the three national legal frameworks shows that there are variations among the countries as regards the provisions protecting privacy in both the constitutions and statutes that apply to video surveillance. After analysing the differences in terms of legislative techniques used by the different countries, we will go on to a comparison of the national rules themselves.

Comparison Concerning the Legislative Techniques

A common characteristic in Sweden, Poland, and Germany is that in all three, in one way or another, the constitution sets limits on video surveillance. However, video surveillance is not directly based on constitutional provisions but on statutes containing either specific provisions aimed regulating video surveillance and/or general data protection rules. The regulation of public and open street video surveillance differs among the countries, as we have seen. In Sweden, the rules applying to the phenomenon of video surveillance are laid down in a specific law about video surveillance, in addition to the general provisions found in data protection legislation. However it has been proposed that the rules be collected in an act on video surveillance covering all non private and non secret video surveillance. Germany as well as Poland lack specific, autonomous legislation regulating video surveillance. In those countries, data protection legislation applies. However, there are differences. The German data protection legislation contains specific provisions regarding video surveillance in publicly accessible places, whereas the Polish legislation does not.

A Brief Comparison of the National Legal Frameworks

In this section we intend to succinctly compare the national frameworks, with focus on the mechanisms aimed at protecting privacy and particularly in regard to the Data Protection Directive. For the German case, only the Data Protection Law at the federal level will be discussed. The comparison is complicated because the countries use different legislative techniques. Instead of focusing on the pieces of legislation themselves, we will focus on the rules. Some protection rules are specifically designed for video surveillance, others are general rules aimed at guaranteeing data protection.

From the overview above we can see that both Sweden and Germany have chosen to enact *specific rules* related to video surveillance in *places which are accessible to the public* (though the Swedish Act on Public Video Surveillance also contains some provisions which apply to non-public places). The scope of the two acts is similar, as both apply to the use of video surveillance undertaken by both private actors and the public sector and both are applicable to all kind of monitoring—for example, analog and digital, sound/video recordings, and video surveillance without storage. Furthermore, both of the acts contain the guarantee that is commonly applied to video surveillance, that is intrinsic to the logic of video surveillance itself—that is, a list, somewhat different in the two cases, of legitimate purposes for video surveillance. Other similar traits include the obligation to provide information whenever an area is being monitored and provisions about the obligation to destroy recorded data. As regards the latter, the requirements diverge somewhat. The Swedish act specifies a delay of one month, while the German act indicates that "the data shall be deleted without delay, if they are no longer needed for the pursued purpose or if the data subject's legitimate interests stand in the way of any further storage". Furthermore, both the acts underline the balance of interests between the purpose served by monitoring and the interest of the individual not to be monitored.

There are, however, a number of differences between the German and the Swedish legal frameworks. The principal one concerns the registration mechanisms. Video surveillance is subject to licensing or notification procedures in Sweden. It is the County Administrative Board that issues licenses or is to be notified. In Germany, the Data Protection Supervisory Authority is to be notified about video surveillance.[72]

The comparison of the legal frameworks of the three countries shows that the transposition of the *data protection* principles applicable to video surveillance—which constitute guarantees for the protection of privacy—is very similar. To begin with, although the three national data protection laws do not explicitly mention sound and image data, the scope of the laws covers both kinds of data. As a general rule, the data protection principles from the EU directive have been incorporated into the three national legal frameworks, although the laws employ somewhat different legislative techniques and do not follow the same design. For example, Article 12(b) of the Directive Concerning the Rectification, Blocking, and Erasure of Data, has been transposed somewhat differently in the three countries: explicitly as a right granted to the data subject (Poland), ambiguously (Germany), and unequivocally (Sweden) as obligations incumbent upon the controller. Another example, but where differences in the transposition of the directive touch on both the form and content of the protection mechanisms, is the right of access to data—provided by Article 12(a) of the directive. The national provisions aimed at transposing this article diverge somewhat, both as regards the scope of the information to which the data subject has access and the intervals at which the data subject may exercise his/her right to obtain information (every six months in Poland, once a year in Sweden, not mentioned in the German act). Furthermore, the waiting period for obtaining information varies. Polish legislation specifies thirty days, and in Sweden it is one month, but the issue is not covered in the German legislation. Finally, they are also variations concerning the cost of requesting information. It is explicitly free of charge in Sweden and Germany, while the Polish law does not address the question.

These examples suggest that the transposition of the directive has been done in somewhat different ways, and that some of the differences may have repercussions on the protecting mechanisms. For instance, whether the request to access data is free of charge or not may influence the inclination of the data subject to make use of the right to access data about him- or herself. Nevertheless, the most significant difference between the three national frameworks regarding *transposing the directive*, one which has a real impact on the guarantees surrounding privacy, stems from the introduction of the abuse-centred model in the Personal Data Act in Sweden in 2007. In practice, this neutralises the protection mechanisms for video surveillance activities in places not open to the public. If we consider the way in which the national frameworks of Sweden, Poland, and Germany regulate *open video surveillance in general*, the most obvious difference concerns the fact that Sweden and Germany, albeit in different ways, have adopted specific provisions on video surveillance in places open to the public. In Poland, only the general data protection rules apply. What do those two observations tell us in terms of Europeanisation? What are the prospects for the regulation of video surveillance?

The Phenomenon of Europeanisation

In the light of the findings reported above, in this section we will reflect on the potential of each of the branches of European law (ECHR law and EU Law) to influence national legal systems. We will also consider the phenomenon of the Europeanisation of rules governing video surveillance. We will then analyse the legal solutions proposed in order to improve the rules regulating video surveillance.

The Potential of Each of the Branches of European Law to Influence National Legal Systems

The process of Europeanisation and the convergence of national legislation are technically more far-reaching when driven by EU law instruments rather than by the ECHR and the case law of the European Court of Human Rights, whose *direct* impact concerns a narrower circle of national legislatures than does EU law. However, concerning the latter, a statement by the European Court of Justice on a case concerning a specific country may *potentially* have consequences on the other member states as well. Indeed, parties to the convention who are suspected of violating the right to privacy in a way that is similar to a state that has already been criticised by the European Court run the risk of being condemned for the same violation. Therefore, as a precautionary measure, member states of the Council of Europe might want to be aware of the conditions specified in the convention that justify interference with human rights guaranteed by the ECHR, and, inter alia, as regards privacy as interpreted by ECtHR case

The Protection of Privacy in the Context of Video Surveillance 119

law, might adapt their legislation for preventive reasons.[73] On this basis, it can be envisaged that the *Köpke* case could have an impact on the Polish legislature, if it wants to ensure compliance of Polish video surveillance regulation with ECtHR case law.[74]

EU law, for its part, is constituted by the compulsory introduction of rules and legal instruments in all the member states of the union. Nevertheless, the level of convergence varies depending on which kind of legal act is used by the EU institutions in order to achieve convergence. Unification, meaning the introduction of identical rules, is achieved by means of *EU regulations*, which have binding legal force in every member state without the state having to pass national legislation. *EU directives* require the member states to achieve a particular result without dictating the means of achieving it. They give states a certain amount of flexibility—depending on how the rules are formulated in the directive—to decide for themselves on the adoption of national rules.

Independent of whether or not video surveillance is regulated in data protection acts at the national level, the Data Protection Directive unquestionably limits the discretion of the national legislatures in matters of the regulation of video surveillance. This is explicitly expressed in the Swedish directive specifying the guidelines that must be followed by the committee of inquiry tasked with examining how to reform video surveillance legislation, which notes that "eventual deviation from the Data Protection Act must be compatible with the Data Protection Directive."[75]

In the meantime, because of the limited field of application of the directive[76], convergence efforts leading to binding rules emanating from the European Union can only be limited, except on a voluntary basis, unless the scope of the rules can be extended. The latter is something that might be a consequence of the entry into force of the Lisbon Treaty.[77]

The Phenomenon of Europeanisation of the Rules Governing Video Surveillance

As mentioned in the introduction to this chapter, the way in which the concept of Europeanisation is understood here is twofold. We conceive the phenomenon as the convergence of the national legal systems of the European countries caused by the influence of EU and ECHR law. In addition, we define it as the integration of European rules, legal institutions, and values into national legal systems.

The Phenomenon of Convergence between the National Legal Systems of the European Countries

The data protection rules derived from Directive 95/46/EC are applicable to the processing of sound and image data in the three countries of interest here; therefore, they are also relevant for video surveillance.[78] Because

Sweden, Poland, and Germany transposed the directive into national legislation in relatively similar ways, there is de facto and de jure convergence of the rules surrounding video surveillance. This convergence will be even greater if the proposed Swedish reform enters into force. Moreover, the three countries recognise the impact of the Data Protection Directive on the way in which they regulate the issue of video surveillance.[79]

However, as noted in the report The Implementation of Directive 95/46/EC on the processing of Sound and Image Data,[80] as well as in Opinion 4/2004 on the Processing of Personal Data by Means of Video Surveillance,[81] and as confirmed by the micro and superficial comparison we conducted here, there are differences in the domestic rules aimed at transposing the EU directive. Those variations may partly be the result of the vagueness of the wording of the directive,[82] and/or because some national legislatures are aware of the special nature of processing in the form of recordings of sound and image data by means of surveillance cameras. In any case, it is not unproblematic. Indeed, "national legal systems might even adopt divergent approaches that could create the danger of 'renationalising', the European legal instruments seeking to create a supranational framework"[83]. This situation would not only have a purely negative economic impact (leading to "higher transaction costs for business and restrictions on market access and free movement"[84]) but would also have an impact on the homogeneity of the level of protection for human rights in the European Union.

Sweden and Germany are actually two countries participating in this tendency to legislate on their own. Interestingly, we can detect a certain convergence in the way in which they understand the issue of the regulation of open video surveillance. Nevertheless, this convergence occurs in a somewhat opposite manner. Indeed, while improved protection of privacy in Germany has consisted of introducing specific rules adapted to the phenomenon of video surveillance in order to supplement the data protection provisions,[85] the Swedish legislature suggests reinforcing the protection of privacy as regards developments in video surveillance by adding, in a new piece of legislation, data protection provisions to the rules enacted in 1998 for regulating video surveillance. Thus, a phenomenon of convergence between the Swedish and German legal frameworks is taking place in spite of the absence of European rules (meaning here EU rules), or perhaps actually because of the absence of a European legal framework.

The Integration of European Rules, Legal Institutes, and Values into National Legal Systems

A certain Europeanisation is discernable concerning the way in which the countries studied here deal with the issue of privacy. Some examples are the reinforcement of the protection of privacy in the Swedish constitution, which was inspired by the European model and, not least, by the values and legal framework expressed in the system of the ECHR.⁸⁶ Another sign of Europeanisation can be detected in the evolution we can observe, inter alia, in the terminology employed by the German Federal Constitutional Court. In recent decisions the court has used the term *fundamental right of information privacy (Grundrecht auf informationelle Privatheit)*, "adapting therefore the constitutional terminology regarding the general right to personality to the international standards".⁸⁷

Legal Solutions in Order to Improve Rules Regulating Video Surveillance

There have been tentative moves by the EU to achieve a uniform application of national measures adopted in order to implement the Data Protection Directive in the area of video surveillance. A good illustration of this is Opinion 4/2004 on the Processing of Personal Data by means of Video Surveillance, adopted by the Article 29 Data Working Party, which is an independent advisory body on data protection and privacy set up under Article 29 of Directive 95/46/EC.⁸⁸ The European Commission, taking into consideration "new challenges for the protection of personal data" caused by "rapid technological developments and globalisation"⁸⁹ promises to "propose legislation in 2011 aimed at revising the legal framework for data protection with the objective of strengthening the EU's stance in protecting the personal data of the individual in the context of all EU policies, including law enforcement and crime prevention, taking into account the specificities of those areas".⁹⁰ For example, the commission proposes to improve the modalities for the exercise of the right of access, rectification, erasure, or blocking of data, by introducing deadlines for responding to individuals or providing that right of access should be ensured free of charge.⁹¹ However, the issue of video surveillance is not the object of specific attention, except concerning a proposal aimed at enhancing the responsibility of the controller and consisting of the introduction into the legal framework of an obligation that this actor conduct data protection assessments in specific cases, including the deployment of video surveillance.⁹² In any case, it will be interesting to see what kind of legal act will be employed to achieve these goals—a directive or a regulation—and if future legislation will be better at taking into account the special characteristics of video surveillance and the need to harmonise the rules that govern it.

The concern for rules better adapted to the phenomenon of video surveillance seems to be more obvious in the framework of the Council of Europe than in the framework of the European Union. After adopting, in 2000, the non-compulsory Guiding Principles for the Protection of Individuals with Regard to the Collection and Processing of Data by Means of Video Surveillance, experts from the European Commission for Democracy through Law, also known as the Venice Commission, recommended in 2007 the adoption of specific regulations at both the international and national levels—"in order to cover the specific issue of video surveillance by public authorities of public areas as a limitation of the right to privacy". Taking into account "the peculiarities and potentialities of the processing, recording and disseminating of video surveillance devices", the Venice Commission makes recommendations which combine rules which can be classified as those typically designed for video surveillance (for example, the recommendation that individuals should be notified that they are being monitored) and for data protection (concerning, inter alia, the principles applicable to the processing of personal data—for example, the principles of fairly and lawfully processing).[93] Furthermore, in 2008 the Parliamentary Assembly of the Council of Europe adopted a resolution on video surveillance of public areas,[94] thereby primarily tackling the issue of video surveillance from a technical point of view, with the aim of containing the perverse effects of technology.

In conclusion, we can note that different reports/surveys show that efficient legal protection of privacy in order to meet the increasing threats posed by technological innovations requires more specific and systematic development.[95] The European legislature has a fundamental role to play as a motor for better harmonisation and the reinforcement of legal mechanisms for protecting privacy against constantly new technological threats. Regardless of the author of the European legal framework, it seems that efficient protection of privacy against interference caused by the use of video surveillance requires a combination of general data protection rules as well as 'tailor-made' ones.

NOTES

1. Venice Commission 2007: 4.
2. Another fundamental right of relevance in this context is the right of free movement. Nonetheless, the present study does not focus on this right.
3. See Bygrave 2004: 327 and the references cited therein.
4. There is no space to develop this issue in this chapter. For a more detailed study, see Bygrave 2004. See also Waldo et al. 2007, especially parts 1 and 2.
5. Federal Constitutional Court (Bundesverfassungsgerichtshof), 15 December 1983, EuGRZ, 1983, 171 ff.
6. Indeed, the Swedish and the French data protection acts refer to personal integrity and private life, respectively. Concerning Germany, if the federal data act makes reference to the personality rights of individuals (*Personlichkeitrecht*), the case law of the Federal Constitutional Court and the states' data protection acts refer to the right of self-determination (see, for example, the data protection acts of Schleswig- Holstein and Brandenburg).
7. See Taylor, N. 2002: 67: "Despite (or because of) the vast literature surrounding 'privacy', it has proved to be a somewhat nebulous concept". Overviews of regional, international, and various national legislations concerning the issue of privacy have been produced, and some of them have a real comparative focus. See, for example, Strömholm 1971 and Bygrave 2004. It is also possible to find reviews focusing on a specific privacy issue such as data protection (Flaherty 1989) or video surveillance at the workplace (Nouwt

et al. 2005). However, scholars do not seem to have addressed the issue of video surveillance in *publicly accessible places* in comparative, in-depth legal analyses.
8. See also the definitions provided by Westin, Gaviscon, Inness in Bygrave 2001, 2004.
9. As some authors have noted, despite its name, *data protection*—which follows data privacy—is not aimed at protecting the data itself but the individuals (Mayer-Schönberger, 1997, 219).
10. The authors refer here to a particularly invasive technology consisting of spyware systems linking employees to their computers with wireless sensors.
11. Landtagsbeschluß zur Videoüberwachung im öffentlichen Raum 08.06.2000.
12. See the cases described in Mayer-Schönberger 2009.
13. BVerfGE 65,1,15 December 1983. Translation suggested by Rouvroy and Poullet 2009.
14. On the supra-European level, two legal documents that constitute catalogues of fundamental human rights contain provisions protecting the *general right to privacy*. These are the 1948 Universal Declaration of Human Rights and the 1966 International Covenant on Civil and Political Rights. In addition, the supra-European level offers legal devices protecting *data privacy*: the 1980 Guidelines Governing the Protection of Privacy and Transborder Flows of Personal Data, which was adopted by the Organization for Economic Co-operation and Development, as well as the United Nations Guidelines Concerning Computerized Personal Data Files, promulgated in 1990.
15. Besides those binding instruments, the Council of Europe has produced nonbinding instruments related to video surveillance such as the Guiding Principles for the Protection of Individuals with Regard to the Collection and Processing of Data by Means of Video Surveillance (2000).
16. This regional organization was founded in 1949 in the aftermath of the Second World War, and today it has forty-seven members. Its goals include promoting human rights and democracy and defending the rule of law.
17. Sweden ratified the convention in 1983, Germany in 1985, and Poland in 2002.
18. Also included is the right to respect for *family life, home, and correspondence.*
19. *Perry v. United Kingdom*, §36.
20. Ibid., §37.
21. See *Peck v. United Kingdom*, §59.
22. *Perry v. United Kingdom.*
23. *Peck v. United Kingdom.*
24. *Karin Köpke v. Germany.*
25. It is interesting to note that despite the ECtHR did not require Germany to set up a legislative framework in order to comply with its positive obligation under Article 8, the German legislature decided to introduced provisions covering this issue in a new Section 32 of the Federal Data Protection Act, which entered into force in September 2009.
26. See, for example, *X and Y v. Netherlands*, §§23, 24 and 27.
27. *Sunday Times v. United Kingdom*, §49.
28. See, for example, *Amann v. Switzerland*, §§55–56.
29. *Karin Köpke v. Germany,*. 11.
30. See *Perry v. United Kingdom*, §45.
31. *Handyside v. United Kingdom*, § 48.
32. See, for instance, *Barthold v. Germany*, §55.
33. *Karin Köpke v. Germany*, 12.

34. The charter, which was proclaimed in December 2000 and became binding with the entry into force of the Treaty of Lisbon in December 2009, applies to EU institutions as well as to member states when they are implementing EU law.
35. See Article 33.
36. The provisions on prior checking concern the processing operations that are likely to present specific risks to the rights and freedoms of data subjects.
37. Article 29 Data Protection Working Party 2004, 6.
38. British Institute of International and Comparative Law 2003, 9–10.
39. Statens offentliga utredningar 2008b: 470–71.
40. Statens offentliga utredningar 2008b: 96.
41. Statens offentliga utredningar 2008b:96. Nevertheless, they do not correspond to enforceable rights (Swedish Parliament. Prop. 1975/76:209, 128).
42. Swedish Parliament. Prop. 2009/10:80, 176.
43. Ibid, 177.
44. Ibid. 180.
45. Ibid, 181.
46. Ibid, 186.
47. Swedish Parliament. Prop. 2009/10:80, 176: "We can assume that Sweden's credibility as signatory to the European convention will increase if the protection of privacy gets clearer support in the constitution".
48. Sweden, which adheres to a dualistic system, ratified the convention in 1952 and incorporated it into domestic law forty-two years later.
49. The act is based in large part on the principles contained in the outdated, initial surveillance legislation, the Act on TV Surveillance (Lag om TV-övervakning 1977:20), which was superseded in 1990 by the Act on Video Surveillance Cameras (Lagen om övervakningskameror m.m. 1990:484).
50. For example, as regards surveillance of post offices, bank offices and shops.
51. The general rule is that image and audio material from public video surveillance of a place to which the public has access must be kept for one month.
52. It replaced the Data Act (Datalagen) of 1973, which was the world's first national data protection legislation, next to a law passed in the state of Hessen in Germany in 1970.
53. Statens offentliga utredningar. 2009:205.
54. The real protective impact of the different mechanisms is, of course, debatable, see Blanc-Gonnet Jonason 2001.
55. See Swedish Ministry of Justice. Dir. 2008:22.
56. Statens offentliga utredningar. 2009:157.
57. Ibid., 165.
58. See, for example, the Constitution of Finland.
59. The wording of article 51(2) shows, by the way, the inspiration of article 8.2 of the ECHR as regards conditions that justify interference.
60. Polish Constitutional Tribunal. 20 April 2004, K 45/02.
61. There are sectoral statutes authorizing video surveillance—as, for example, the Law of 6 April 1990 on the Police, the Law of 29 August 1997 on Municipal Police, and the Law of 20 March 2009 on Mass Events Security.
62. The Data Protection Directive was transposed into Polish law when Poland was a EU candidate country; Poland became a EU member in May 2004.
63. Polish Inspector General for Personal Data Protection 2004: 141–42.
64. See, for example, the decision of the Constitutional Federal Court on 23 February 2007 mentioned below.
65. See Weichert 2001.
66. BVerfG, 1 BvR 2368/06, http://www.bverfg.de/entscheidungen/rk20070223_1bvr236806.html

67. The complainant was a lawyer who passed by the monitored area on a daily basis.
68. It is also regulated by sectoral rules—for instance, the Federal Police Act and the Federal Criminal Police Office Act.
69. The law, from 2001, has been revised several times (most recently in 2009), and replaced the first German data protection law from 1977 (revised 1990).
70. Weichert 2001.
71. The act includes a section dedicated to data processing by public bodies (Part II) and a part dedicated to data processing by private bodies and commercial enterprises under public law (Part III).
72. The obligatory registration (Section 4d) does not apply when a data protection official is appointed by the party that controls the data.
73. In general, the Swedish legislature, before enacting a law that might infringe on a right guaranteed by the ECHR, pays attention to the text of the convention as well as to the way in which the ECtHR has interpreted the relevant provision. It is also interesting to note that the Swedish National Court of Administration (Domstolsverket) publishes in its newsletter directed, inter alia, to the Swedish courts, recent case law originating from the ECtHR.
74. *Karin Köpke v. Germany*, Decision as to the Admissibility of Application no. 420/07.
75. Swedish Ministry of Justice. Dir. 2008:22. In the preparatory work of the current Act on General Video Surveillance, the impact of the Data Protection Directive, which was under transposition at that time, was taken into consideration. See Prop. 1997/98:64.
76. The directive does not apply to "processing operations concerning public security, defence, state security and the activities of the state in areas of criminal law" (Article 3.2).
77. See European Commission 2010a: 4. A new legal basis, Article 16 of the Treaty on the Functioning of the European union, allows the EU to have a single legal instrument for regulating data protection, including the areas of police cooperation and judicial cooperation in criminal matters.
78. In Sweden, thus far, in principle they are only relevant for video surveillance in non-public places. If the proposed reform of the Act on Public Video Surveillance is adopted, it will extend the application of some of the data protection rules to video surveillance in public places.
79. For Poland, see Polish Inspector General for Personal Data Protection 2004; for Germany, see, for example, Weichert 2000; for Sweden, see Swedish Ministry of Justice. Dir. 2008:22.
80. British Institute of International and Comparative Law 2003.
81. Article 29 Data Protection Working Party 2004.
82. See European Commission 2010b: 29.
83. British Institute of International and Comparative Law 2003: 70.
84. Ibid.
85. This development has taken place at the federal as well as state level, as the Bavarian case illustrates. It is interesting to note that the introduction of specific rules for regulating video surveillance has been the object of an internal initiative as regards *video surveillance to places open to the public*, while the introduction of federal rules concerning *video surveillance at workplaces* has followed a complaint handled by the ECtHR in the *Köpke* case (although the complaint was considered non-admissible by the court).
86. See above under the Swedish legal framework.
87. See Der Bayerische Landesbeauftragte für den Datenschutz. 2008. 23. Tätigkeitsbericht, 12.

88. We can notice that in the final report from January 2010, the primary proposal made in order to achieve a greater harmonisation of the data protection rules consists is to rely on this working group: "Although its opinions etc. are not binding, it has the expertise, and the direct links with national practices, to be able to formulate interpretations and manners of application of the provisions of the Directive" (European Commission 2010b: 40).
89. European Commission 2010a: 2.
90. Ibid. 18.
91. Ibid. 8.
92. Ibid. 12.
93. Venice Commission 2007.
94. Council of Europe, Parliamentary Assembly 2008.
95. In this context, surveillance/video surveillance is only one of several issues which require a more efficient legal framework. Other challenges for the protection of personal data are caused by, for instance, technical phenomena such as the Internet or special new technologies such as biometrics and radio frequency identification.

REFERENCES

Bibliography

Banisar, David. 2000. *Privacy and Human Rights: An International Survey of Privacy Laws and Developments.* Washington D.C.: Electronic Privacy Information Center. Accessed 11 July 2008 at www.privacyinternational.org/survey/.

Blanc-Gonnet Jonason, Patricia. 2001. "Protection de la vie privée et transparence à l'épreuve de l'informatique, droit français, droit suédois et directive 95/46/CE du Parlement européen et du Conseil du 24 octobre 1995". PhD diss., University of Paris XII.

Bygrave, Lee A. 2001. "The Place of Privacy in Data Protection Law." *University of New South Wales Law Journal* 277. Available at http://www.austlii.edu.au/au/journals/UNSWLJ/2001/6.html#Heading13

Bygrave, Lee A. 2004. "Privacy Protection in a Global Context—A Comparative Overview." *Scandinavian Studies in Law* 47: 319–48.

Clayton, Richard, and Hugh Tomlinson. 2000. *The Law of Human Rights.* Oxford: Oxford University Press.

Flaherty, David H. 1989. *Protecting Privacy in Surveillance Societies: The Federal Republic of Germany, Sweden, France, Canada, and the United States.* Chapel Hill: University of North Carolina Press.

Gavison, Ruth. 1980. "Privacy and the Limits of Law." *Yale Law Journal* 89: 421–71.

Hornung, Gerrit, and Christoph Schnabel. 2009. "Data Protection in Germany I: The Population Census Decision and the Right to Informational Self-determination." *Computer Law and Security Report* 25(1): 84–88.

Inness, Julie C. 1992. *Privacy, Intimacy and Isolation.* Oxford: Oxford University Press.

Krizsán, Andrea. 2001. *Ethnic Monitoring and Data Protection: The European Context.* Budapest: Central European University Press.

Mayer-Schönberger, Viktor. 1997. "Generational Development of Data Protection in Europe." In *Technology and Privacy: The New Landscape*, ed. Agre, Philip E. and Marc Rotenberg. Cambridge, MA: The MIT Press. 219–242.

Mayer-Schönberger, Viktor. 2009. *Delete: The Virtue of Forgetting in the Digital Age.* Princeton, N.J.: Princeton University Press.

Nouwt, Sjaak, Berend R. De Vries, and Corien Prins, eds. 2005. *Reasonable Expectations of Privacy? Eleven Country Reports on Camera Surveillance and Workplace Privacy.* The Hague: TMC Asser Press.

Prost, Antoine. 1987. "Frontières et espaces du privé" In *Histoire de la vie privée*, ed. Ariès, Philippe and Georges Duby. Paris: Éditions du Seuil, Tome 5.

Regan, Priscilla M. 2002. "Privacy as a Common Good in the Digital World." *Information, Communication and Society* 5(3): 382–405.

Ross, Jacqueline. 2005. "Germany's Federal Constitutional Court and the Regulation of GPS surveillance." *German Law Journal* 6: 1805–12. Available at http://www.germanlawjournal.com/index.php?pageID=11&artID=678.

Rouvroy, Antoinette, and Yves Poullet. 2009. "The Right to Informational Self-determination and the Value of Self-development: Reassessing the Value of Privacy for Democracy." In *Reinventing Data Protection?*, ed. Gutwirth, Serge, Yves Poullet, Paul De Hert, Cécile de Terwangne, and Sjaak Nouwt. Berlin: Springer-Verlag, 45–76.

Strömholm, Stig. 1971. "Integritetsskyddet—Ett försök till internationell lägesbestämning." *Svensk Juristtidning,* 56 : 695–736.

Taylor, Nick W. 2002. "State Surveillance and the Right to Privacy." *Surveillance and Society* 1(1): 66–85.

Tsakyrakis, Stavros. 2008. *Proportionality: An Assault on Human Rights?* Jean Monnet Working Paper 09/08. Available at http://centers.law.nyu.edu/jeanmonnet/papers/08/080901.html

Waldo, James, Herbert S. Lin, and Lynette I. Millett, eds. 2007. *Engaging Privacy and Information Technology in a Digital Age.* Washington, D.C.: National Academies Press.

Warren, Samuel, and Louis Brandeis. 1890. "The Right to Privacy." *Harvard Law Review* 4(5): 193–220.

Weichert, Thilo. 2000. *Aktuelle Fragen zur Videoüberwachung.* Available at http://www.datenschutzzentrum.de/video/videosec.htm.

Weichert, Thilo. 2001. *Private Videoüberwachung und Datenschutzrecht.* Available at http://www.datenschutzzentrum.de/video/videpriv.htm.

Westin, Alan. 1970. *Privacy and Freedom.* New York: Atheneum.

Whitman, James Q. 2004. "The Two Western Cultures of Privacy: Dignity versus Liberty." *Yale Law Journal* 113:1151–1221. Available at http://digitalcommons.law.yale.edu/fss_papers/649.

Young, Raymond. 2006. *English, French and German Comparative Law.* London: Cavendish.

Legislative Reports and Other Materials

Article 29 Data Protection Working Party. 2004. Opinion 4/2004 on the Processing of Personal Data by means of Video Surveillance.

British Institute of International and Comparative Law. 2003. The Implementation of Directive 95/46/EC to the Processing of Sound and Image Data.

Classen, Claus Dieter, Ilona Stolpe, and Manfred Bornewasser. 2008. Videoüberwachung öffentlicher Strassen und Plätze.Frankfurt:Verlag für Polizeiwissenschaft

Council of Europe, Parliamentary Assembly. 2008. *Video Surveillance of Public Areas.*Resolution 1604 (2008).

European Commission. 2010a. *Communication from the Commission, the European Parliament, the Council, the Economic and Social Committee and the Committee of the Regions: A Comprehensive Approach on Personal Data Protection*

in the European Union. Brussels: European Commission. Available at http://ec.europa.eu/justice/news/consulting_public/0006/com_2010_609_en.pdf.
European Commission. 2010b. *Comparative Study on Different Approaches to New Privacy Challenges, in Particular in the Light of Technological Developments*. Brussels: European Commission. Available at http://ec.europa.eu/justice/policies/privacy/docs/studies/new_privacy_challenges/final_report_en.pdf.
Polish Inspector General for Personal Data Protection. 2004. Activity report.
Statens offentliga utredningar. 2008a. En reformerad grundlag Report no. 2008:125. Available at http://www.regeringen.se/sb/d/10025/a/117744.
Statens offentliga utredningar. 2008b. Skyddet för den personliga integriteten. Report no. SOU 2008:3. Available at http://www.regeringen.se/content/1/c6/09/63/73/61a22251.pdf.
Statens offentliga utredningar. 2009. En ny kameraövervakningslag. Report no. 2009:87. Available at http://www.regeringen.se/sb/d/108/a/134670.
Swedish Ministry of Justice. Dir. 2008:22. Översyn av lagen om allmän övervakning m.m.
Swedish Parliament. Prop. 1975/76:209. Om ändring i Regeringsformen.
Swedish Parliament. Prop. 1997/98:64. Lag om allmän kameraövervakning.
The Schleswig-Holstein Landtag. 2000. Landtagsbeschluß zur Videoüberwachung im öffentlichen Raum 08.06.2000. Available at https://www.datenschutzzentrum.de/video/video_lt.htm.
Venice Commission. 2007. CDL-AD (2007) 014, Opinion on Video Surveillance in Public Places by Public Authorities and the Protection of Human Rights.

Judicial Cases

European Court of Human Rights

Amann v. Switzerland. Application no. 27798/95, judgment 16.02.2000.
Barthold v. Germany. Application no. 8734/79, judgment 25.03.1985.
Handyside v. United Kingdom. Application no. 5493/72, judgment 07.12.1976.
Karin Köpke v. Germany. Decision as to the admissibility of application no. 420/07, 05.10.2010.
Peck v. United Kingdom. Application no. 44647/98, judgment 28.01.2003.
Perry v. United Kingdom. Application no. 63737/00, judgment 17.10.2003.
Sunday Times v. United Kingdom, application no. 6538/74, judgment 26.04.1979.
X and Y v. Netherlands. Application no. 8978/80, judgment 26.03.1985.

Domestic Courts

German Federal Constitutional Court (BVerfG). Ruling 65,1,15 of December 1983.
German Federal Constitutional Court (BVerfG). Ruling 1 BvR 2368/06 of 23 February 2007. Available at http://www.bverfg.de/entscheidungen/rk20070223_1bvr236806.html.
Polish Constitutional Tribunal. Judgment of 20 April 2004, K 45/02. Available at http://www.trybunal.gov.pl/eng/summaries/documents/K_45_02_GB.pdf.

Part II
Case Studies

5 Video Surveillance and the Question of Trust

Wojciech Szrubka

Do the Poles like living under camera surveillance? The simple answer appears to be a resounding *yes*. Video surveillance cameras seem ubiquitous in Poland—not just in the big, urban agglomerations like the capital Warsaw, but also in towns and villages; and their number seems to be constantly on the rise. This development does not appear to be simply a result of an elite-driven, top-down campaign; in Poland, video surveillance of public places seems to enjoy tremendous popular support (Gniadek 2009).

What is the driving force behind this public enthusiasm towards modern surveillance measures? A couple of answers seem close at hand: the Poles have probably not yet abandoned their expectation of the state as their 'big protector'—an expectation created at the time of communism, when excessive state control over almost every aspect of social life was motivated by the stated intention to 'protect' citizens from danger. Even though the role of the state as a benevolent 'big brother' has been discredited and ridiculed in the public discourse, it is perhaps not far-fetched to suppose that some expectations of the state, especially its duty to 'protect', are still deeply entrenched in the public mind and have effectively survived communism. The strong public support for camera surveillance could therefore be explained by this expectation being so strong that it overrides any fears one could have about excessive state control and invigilation.

Another, similar, reason could perhaps be found in the Poles' experience of state power over the past five decades. The communist regime in Poland, while sharing many oppressive traits with its Soviet or East German counterparts, may in fact have been, if not more benevolent, then perhaps less well-organised and therefore less effective than the others in implementing its totalitarian agenda (Garton Ash 1999). Unlike the East Germans, or even some Western European citizens, the Poles are probably not too afraid of their state simply because they have rarely experienced its sheer effectiveness and severity. When given the choice of being free from state supervision or being put under limited surveillance with a vague promise of safety, many Poles are perhaps inclined to choose the latter because they do not see any immediate risk to their privacy or freedom.

Both explanations sound reasonable and may well be proved correct by subsequent research. The aim of the present article is to discuss the viability of a third possible explanation of the seemingly broad popular support for camera surveillance in Poland. The idea being put forward here could be summarised as follows: the popularity of video surveillance may under certain circumstances be conditioned by a low level of social capital—that is, weak interpersonal trust. People do not trust each other and are therefore willing to accept video surveillance as a way of making sure that everyone is fulfilling his or her part of the 'societal deal'. Therefore they accept video surveillance not so much in order to grant extraordinary powers to the police but to ensure fellow citizens do not commit crimes or damage public (or private) property.

Discussing the viability of an explanatory model relying on distrust is crucial here and is the main purpose of this chapter. This is different than arguing for the validity of a certain explanatory model of the popularity of video surveillance in Poland; such a project would require a more complex research approach, involving a larger number of in-depth case studies, as well as more quantitatively oriented research, where the correlation of interpersonal trust with public support for surveillance (including video surveillance) could be detected and measured. Here, I first discuss on a theoretical level in what way a deficit of trust could create increased demand for monitoring. Thereafter, I review a number of cases of video surveillance in which a deficit of interpersonal trust appears to have played a substantial role. The cases are briefly presented and the role of trust (and distrust) in the way video surveillance is being utilised is discussed. The function of the cases is thus mainly to illustrate the expected effects and mechanisms that were discussed in the theoretical part.

Having said that, the choice of Poland as a reservoir of cases is no coincidence—like many of its Eastern European counterparts, its social capital, measured as levels of interpersonal trust, is low (see Tables 5.1 and 5.2 in the chapter appendix). If there were to be a systematic study of the impact of distrust on the popularity of video surveillance, Poland would no doubt constitute an intriguing case. (More on the subject of social capital later on.)

In other words, the hypothesis is that broad public support for surveillance measures in, for example, Poland may be a consequence of a severe deficit of interpersonal trust in that particular society. The term *video surveillance* is used here in the broadest possible sense, and includes public as well as private video camera supervision and some examples in between. Our focus is the use of video equipment in order to gather hard evidence (either incriminating or corroborating), as well as the use of cameras in situational crime control. Thus, a camera located inside a store or on the ceiling of a hotel lobby will be equally as interesting as a camera overlooking a town square or a tiny device inside an ambulance. All of these devices serve the purpose of amassing evidence of some sort or are tools in direct (situational) crime prevention.

Studies connecting trust to popular support for video surveillance appear to be almost non-existent, which is rather surprising, considering the size of the role surveillance cameras have in combating crime, and the intuitive links one could draw between camera density and fear of crime—the latter being a manifestation of a deficit in interpersonal trust. A rare example of such research is a short study by Neyland titled *Privacy, Surveillance and Public Trust* (2006). As the title suggests, the focus of Neyland's paper is the issue of privacy and public trust towards the institutions deploying video surveillance (whether public or private). This work, while related to our own here, does not touch upon the issue of interpersonal trust as a possible driver of the growing use of closed circuit television (CCTV).

THE BOTTOM LINE: IS ALL VIDEO SURVEILLANCE ABOUT DISTRUST?

A simple question begs an answer right at the start: Is not all video surveillance ultimately motivated by distrust? After all, in a trustful society, a society without malice and insincerity, there would surely be much less need for surveillance measures. Surveillance, to be sure, would probably still be utilized, but its purposes would be different: cameras could be deployed to monitor the traffic situation on a motorway, or to feed information to a train operator on whether the train doors have been properly shut and no passenger has been squeezed in the train exits. Gathering information for the purposes of targeted marketing or quality control would be another possible application of video surveillance.

As soon as cameras are used to prevent crime or prosecute law offenders, the question of trust naturally enters the equation. A question that quite obviously arises here is, in what way is the situation different in societies marked by trust deficit, from those that possess large reservoirs of social capital? My simple answer to this would be that there is *no difference* concerning the underlying mechanism—trust or distrust is very likely a major driving force in all national settings where video surveillance is being utilised. As in many other contexts, the difference will be first and foremost a matter of degree, meaning that the factor under consideration (here, *distrust*) will make its special mark on the way in which video surveillance is utilised. That is to say, the distrust which characterises societal relations is likely to influence the underlying functions (purposes) of video surveillance and stronger popular support for video surveillance can be expected. As a result of higher levels of distrust, we can also expect a significantly larger number of video cameras to be deployed.

A TRUSTFUL SOCIETY: SOCIAL CAPITAL

What is the core difference between an individual possessing huge amounts of social capital and one who does not? From the scholarly literature on the subject, we learn that the former individual will possess a dense network of contacts and he or she will benefit from these social networks in many different ways (Putnam 2000: chap. 1). These contacts will help him or her to find a new job or get assistance in times of trouble. This individual will develop trust towards the people forming his or her network of contacts. He or she will trust their good intentions, and will therefore not hesitate to offer them his or her helping hand (or even donate an organ) when they are in need (Putnam 2000). If the individual rich in social capital is a part of an extended and inclusive network of trust, there are important fringe benefits for the society as a whole. The trust is no longer limited to the individuals with which he or she has a personal relationship; it becomes universal, and extends to the surrounding people in general. Moreover, it extends to the public institutions and those running them, which thus strengthens and legitimises the state (the latter applies, of course, only to democratic states, which derive their authority from the popular vote). The latter fact is quite significant and means that the trust that develops is not strictly related to specific persons' past records of fulfilled agreements, nor is it strictly connected to any specific cases of returned favours. Rather, its basis is a tacit agreement of *reciprocity*—I do you a favour today and you (or somebody else) will return the favour later. Reciprocity becomes a general value underlying people's actions. To use the term coined by March and Olsen, *reciprocity*, being an active, contributing member of society becomes a part of the "logic of appropriateness" (March and Olsen 2004).

One more remark concerning the consequences of interpersonal trust is perhaps warranted. The theory of social capital does not usually distinguish between interpersonal trust on the one hand and trust between citizens and public institutions/officials on the other. One thing follows naturally from the other; interpersonal trust permeates all social relations. Since social capital (or its absence) is viewed here as the chief independent variable in explaining the popularity of video surveillance, I will also assume that its presence has positive consequences for the popularity of public institutions and the reciprocal trust between citizens' and public officials.

THE ISSUE AT STAKE: THE IMPORTANCE OF TRUST AND SYMPTOMS OF TRUST DEFICIT

The advantages of strong interpersonal trust for a society are obvious, and I do not need to elaborate on them here. I do, however, need to reflect on the exact functions that trust fulfils. *Trust* is a needed component in the social machinery, and its lack is likely to manifest itself in certain ways. There are a number of approaches to studying the functions of *trust*, just as there are several competing perspectives on the role of institutions and their influence on society (Misztal 1996). The reflection on the functions of trust summarised here is therefore but one of several perspectives that can be found in the academic literature. I will follow the perspective of *rational choice* for one simple reason: social capital theorists argue for the usefulness of interpersonal trust from exactly this position, and their argument involves resolving the dilemmas of co-operation, particularly the 'prisoners' dilemma', and the related problem of 'free-riding'. The following discussion will shed further light on the crucial question I am posing—namely, how can a lack of interpersonal trust make the use of video surveillance desirable?

FUNCTIONS OF TRUST

Rational choice theorists are, by virtue of the very axioms that form the basis of their theory, preoccupied with the dilemmas and problems of collective action. The approach is characterised by a methodological individualism, which assumes that "any participation in collective action can be explained by models of rational individual action, where rationality is understood in utilitarian terms as a matter of satisfying the individual's preference, and consists of choosing that action that is most likely to produce the highest utility for the actor" (Misztal 1996: 77). The crucial challenge in a world populated by individuals (and other actors) pursuing their own interests (and who only possess limited information about the other actors' plans and intentions) is to make *co-operation*, rather than *discord*, the most beneficial mode of behaviour. Cooperation is not viewed here as inherently superior in absolute moral terms; rather, it is preferable because it is optimal from the perspective of each actor, given that all (or at least a substantial number of) other actors choose to cooperate simultaneously. Institutions then have to be constructed in a way that facilitates cooperation and discourages defection. Put differently, the crucial task for institutions from a rational choice perspective is to overcome the problem of free-riding; even though each of the actors may well be aware that there is a collective, long-term interest in upholding contracts and cooperating, the allure of an easy gain by defecting may be sufficiently tempting. The defection of some will have a double effect. First, it will disrupt the regulated distribution of a common resource (like fresh water or pasture land), giving unjustified advantage to some, and placing a disproportionate burden on others for developing a resource that, once produced, will become freely available to all parties (unencrypted terrestrial public service television and radio are good examples of this). The second effect will arguably be more serious than the first: it will undermine trust. The participating actors may be perfectly honest and have a genuine will to cooperate, but once they perceive the other participants as free riders they will very likely defect too. If the original agreement concerns the use of a scarce resource, then, and given that there is distrust between the participating actors, it would be rational for each party to snatch as large a chunk of it as possible before it is depleted. As for the production of public goods, the motivation for their continued generation will be eroded. Cooperation will cease and, ultimately, every actor will be worse off. The proponents of the theories of social capital and rational choice seem markedly similar in their description of the benefits of trust: they view it in a utilitarian way. Trust is the 'lubricant' of cooperation (Arrow 1974: 23). Since public institutions depend on cooperation between public officials and citizens, they will only function well if there is enough of this lubricant. This will depend, among other things, on how trustworthy (uncorrupted) these public officials are perceived to be (Putnam 1996, 2000). Even though social capital theorists appear to disagree on how the social capital of interpersonal trust is created in the first place, they do agree that its lack will be detrimental to the state, its institutions, and society as a whole. It will exacerbate the problem of free-riding and the 'tragedy of the commons'; it will furthermore make the outcome of economic transactions suboptimal by raising costs.[1]

The problem of free-riding or the tragedy of the commons has traditionally been resolved in two different ways (Rothstein 2008). The first solution entails a pessimistic view of the human nature and may be called *Hobbesian*. It envisages a scenario, in which naturally selfish individuals will sooner or later go to war and annihilate each other unless somebody stops them. That *somebody* must assume the role of a harsh ruler, who will closely monitor the actions of his subjects, curb their destructive (selfish) behaviour, and force them to follow a constructive set of common rules. In more practical terms, the solution could entail a strong state with an effective and watchful police force and other control and sanctioning mechanisms.

The second solution involves creating institutions and fostering trust between actors. Locally created institutions for managing the use of fragile, limited natural resources have proved to be an effective way of overcoming the tragedy of the commons (Ostrom 1990). The institutions foster regular contact between actors and, most importantly of all, force the actors to explain and justify their actions in moral, unselfish terms. The actors could thus deliberate with each other on their common good, which strengthens their mutual trust still further. In more general terms, strong interpersonal trust and, by the same token, trust towards public institutions will be an effective solution to free-riding (and the tragedy of the commons). This is a preferable solution to the first, because it eliminates, or greatly reduces, the cost of monitoring (Coleman 1990: 306–10), and monitoring can never be made perfect.

We can safely assume that contracting parties will probably never rely purely on either trust or monitoring and sanctions but instead on a combination of both. It is conceivable that there will be a continuum between the two extremes, with high trust and minimal monitoring and sanctioning at one end of the scale and minimal trust and maximum monitoring and sanctioning at the other.

If we apply these findings to a society marked by low social capital, the pervasive lack of trust could be expected to leave its mark on the way this society works. In societal deals and contracts, the balance between the Hobbesian solution (sanctions and monitoring) and the institutional one (trust) will probably give more emphasis to the former. Societal co-operation, understood here as the state's dealings with the citizens as well as dealings between the citizens excluding the state, will likely be marked by a persistent, obsessive urge to eliminate fraud (free-riding), and the tools used to achieve this would likely be *policing*, *monitoring*, and *sanctions* (Sztompka, 1999: 116–18). It is here that the use of video surveillance equipment can be seen as a vital element in the 'arsenal' of monitoring tools.

SOCIAL CAPITAL IN POLAND

Social capital, understood as interpersonal trust, is famously low in the countries of the former Eastern bloc, and this was still true for Poland at the end of the first decade of the twenty-first century—almost two decades after the era of state socialism ended there. For the purpose of this article, cumulative data has been compiled from the two latest rounds of the European Social Survey (2006 and 2008; see the chapter appendix). Tables 5.1 and 5.2 show cumulative results from Poland, Sweden, and Germany. The results reveal substantial differences in the level of interpersonal trust between Sweden on one side and the two remaining countries on the other. Sweden appears to have the highest scores and, while Poland and Germany are fairly close together, Poland still stands out clearly as the country with the lowest reservoir of interpersonal trust (see the appendix).

THE USES AND FUNCTIONS OF VIDEO SURVEILLANCE—WHAT SHOULD WE EXPECT?

Trust, or lack of trust, as understood by social capital theorists is a feature pertaining to individuals. When people do not find each other trustworthy, the distrust will likely manifest itself as assuming the worst about the intentions of others; in this scenario, people likely will:

- use (the threat of) sanctions to minimise their risk
- try to extensively monitor each other's actions, believing that lack of monitoring will open the floodgates to free-riding and fraud

Stemming free-riding by placing sanctions on cheating individuals requires the obtaining of hard evidence, and the demand for such evidence—either incriminating or proving somebody's innocence—will be high. The expectation will also be that making people aware that they are being monitored is the only feasible way of ensuring that they behave decently.

EMPIRICAL CASES OF VIDEO SURVEILLANCE IN POLAND

In this section I will be reviewing a number of empirical cases of the use of video surveillance. The obvious question one has to pose at this point is that concerning the representativeness of these empirical examples. The cases may look interesting, but how much do they really say about the 'whole picture'? At this point, I would like to remind the reader about the statement made at the beginning of this chapter: this is intended as a *pilot study*, the sole purpose of which is to discuss and illustrate the importance of trust (and distrust) as a driving force behind the popularity of video surveillance. The reviewed cases should thus be viewed primarily as illustrations of theories. The question of whether distrust is the main driving force behind the dynamic growth of video surveillance in a particular country—in this case Poland—must be answered in a separate study. Nevertheless, it may be possible to see the reviewed cases as strong evidence of this, and too significant to be dismissed out of hand. The possible significance of the cases stems, as I see it, from two sources.

First, two of them are too extensive to be ignored, and by *extensive* I mean the large number of cameras deployed, and also the number of people being subjected to their supervision. The municipal video surveillance in Warsaw, for instance, involves several hundred cameras supervising hundreds of thousands of people every day, and is by this fact alone difficult to dismiss, even though this particular case only highlights the situation in the Polish capital city. The use of video surveillance equipment in driving test vehicles concerns one particular and, one could argue, *limited* application area. The fact remains, however, that there are hundreds of these vehicles all over the country. Moreover, these cameras were deemed to play such an important role that their deployment has become a legal requirement. This case, therefore, also weighs heavily enough not to be dismissed.

Two other cases of surveillance are on a smaller scale but possess another quality that makes them significant: they are odd in a sense that they would probably be unthinkable in many other comparable countries. The most salient example of such a case could perhaps be the use of camera surveillance in medical ambulances. The mere fact that such an idea has surfaced should arouse questions as to what kind of wider societal phenomena are in play; and the evidence points strongly towards *distrust* as the main explanation. The aforementioned cases could thus be seen as evidence that Poland constitutes a good candidate for a future study of possible links between levels of distrust and the popularity of video surveillance. In this sense, the choice of Poland as a case study was no coincidence.

Passive Surveillance

Video surveillance most likely fulfils several functions, and I will review the most important of these later on. For now, it is important to ask what kind of video surveillance would best fulfil the role of amassing evidence. Naturally, we would look first to 'passive' video recording, which means that no one is monitoring the video recordings in real time, and it is never used to stop a crime in progress. In fact, nobody will ever see the footage unless there is a need to reconstruct the events in connection with a criminal investigation. The device could be concealed, but this is not really necessary, for what matters is only that the material is gathered and stored for some definite (or indefinite) amount of time and is ready to be used should there emerge a need to do so. Moreover, we would expect that, in such cases, gathering evidence would be the stated goal of video surveillance, and that its role would be that of an 'insurance policy'. Two instances of such video surveillance will be presented here, both of which have been, or are still, in use in Poland.

Case 1: Video Surveillance in Emergency Ambulances

In September 2006, electronic and print media in Poland reported that fourteen newly purchased ambulances, due to be put in service in the city of Warsaw, would all be equipped with automatic video surveillance cameras. Patients' journeys to the hospital, their struggle for life, and the actions of the medical staff would all now be documented and the records saved for some time. It is not totally clear what triggered this move on the part of Warsaw Municipal Emergency Services, but it is probably not far-fetched to assume that it had something to do with the widely reported revelation from the city of Łódź that emergency ambulance personnel had allegedly murdered some of their patients in order to sell their corpses to undertakers (Łowcy Skór 2012). By installing video equipment, the management of the emergency services wanted to make sure that there would be enough evidence to free their staff from any suspicion of crime or inappropriate conduct should some accusations or allegations of this kind emerge. In the words of Krzysztof Rolirad, the Manager of Warsaw Emergency Services, "People are getting more and more litigious. Legal processes drag on for years. In case our employees are accused of theft or negligence in fulfilling their duties, such a recording can constitute an important proof" (Pochrzęst 2006).

The same article reports that the medical doctors—in whose interest the video equipment had been installed—were the most vocal opponents of the move, and their protests eventually led to the removal of the cameras. Two reasons were given for their opposition: the first had to do with the patients' right to privacy in the face of sickness and death; the second one was no doubt connected to the perceived lack of trust in them on the part of the patients: "We use the cameras only in order to improve the quality of care and security of patients, for instance in the intensive care units. The cameras in the ambulances surely do not serve such a purpose. You cannot assume that medical personnel are the enemy of the patients and that you always have to watch their hands" (Konstanty Radziwiłł, chairman of the Chief Physicians' Chamber, quoted in Pochrzęst 2007). Of what little is known about the opinion of the potential patients on the use of cameras, two voices are perhaps worth quoting:

> My sister had once lost her ring in an ambulance. We had no proof that it had been stolen. (Michał Karaś, a twenty-year-old man from Warsaw, quoted in Pochrzęst 2006)

> What are the doctors afraid of? If their conscience is clean and if they only help their patients, then they should not care about the recordings. (Agnieszka Gołębiewska, a Warsaw woman, quoted in Pochrzęst, 2006)

This case seems to be a good example of the phenomenon being focused on here: lack of trust; the original purpose of the cameras was to stem a potential wave of litigation from distrustful patients. The function of the cameras was to secure evidence, as proof of innocence or guilt on the part of the medical staff. The medical personnel had no capital of trust and 'needed' to be monitored; they were potential suspects of thefts or, worse, negligence (or even pre-meditated actions) leading to somebody's death. And precisely because the presence of cameras implicated open distrust, the move was deemed unacceptable by the medical doctors. The doctors could tolerate video equipment serving other purposes (quality control and patient safety) but not if its sole purpose was to free them from suspicions of crime. The opposition against the move on the part of the medical staff eventually led to the removal of the cameras.

Case 2: Video Surveillance of Driving Tests

In Poland, organising and carrying out driving tests is the responsibility of the regions (*województwa*). There are sixteen regions with limited local government prerogatives, and in each region there is at least one (but often more than one) Regional Road Traffic Board office, whose task it is to execute driving tests and issue driving licenses. In April 2006, the Polish Ministry of Transport and Construction ordered that all driving test cars had to be equipped with video cameras to record events on the road in front of the car as well as the sounds inside it. According to the ministry, the cameras' main purpose is to counter corruption. They are supposed to verify the completeness of the exam and whether the chosen route and length of exam are appropriate (Egzamin na prawo jazdy z kamerą i mikrofonem 2009).

The use of video surveillance in connection with driving tests has gained some media coverage, which also gives us an idea as to the practical use of this new technical facility. Apart from protecting the candidates from the arbitrariness of inspectors, and serving as material proof should there be any subsequent disputes, the recordings can, theoretically, document candidates' mistakes should they fail to pass the exam. But, above all, they should prevent corruption.

This case is another one that illustrates the use of cameras as evidence-gathering equipment. The recordings should constitute material proof that the exam has been carried out correctly. They should prevent corruption (bribery) by their mere presence and, most important of all, be the proof of guilt or innocence in case of any controversies. The fact that the need to equip the cars with cameras has been motivated by distrust is self-evident. The capital of trust of the Regional Road Traffic Authority was insufficient, and had to be compensated for by increased monitoring, which was achieved by installing the cameras.

PUBLIC VIDEO SURVEILLANCE WITH MIXED PURPOSES

The two cases just described represent very strongly the phenomenon which is the main subject of the present chapter: it is evident that the use of video camera surveillance here is heavily focused on gathering evidence of crime, and has been propelled by the distrust between the involved actors. Since at least one category of actors in each case fears being 'cheated' by other actors on whom they depend, they see *monitoring*—and in this case the term obtains an almost literal meaning—as a vital way to secure their interests.

These two cases, however interesting, represent only a small fraction of total video surveillance in Poland. Is there any evidence that the expansion of the more mainstream video surveillance of public spaces—carried out by, for instance, the Polish police—could be driven by a severe lack of trust? In a moment I will argue that this is indeed the case, and will illustrate it with at least three examples. Here however, we are concerned with the kind of video surveillance that probably lacks a clear single goal and can be assumed to fulfil several functions at once. In order to make my case, I will therefore have to present clear indications that some particular goals of video surveillance—namely, those related to gathering evidence of crime—take prominence over others.

First of all, it would be instructive to find out information about the officially stated goals of this kind of video surveillance. What is the official justification for the use of video cameras to monitor public spaces? A short excerpt from a Polish law, regulating the work of the police and city guards, may perhaps provide us with a clue. In fulfilling their task of securing order and peace in public places, the city guard has a right to use "technical means of recording the sequence of events". In particular, these technical means should be utilised in order to:

1. secure the evidence of a crime or offence
2. prevent disturbances to the peace and order in public places

3. protect public buildings and public facilities (Internetowy System Aktów Prawnych 2012)

If the numerical order of the tasks has anything to tell us as to the importance of each of them, *securing evidence* is perhaps seen as playing a more prominent role, as it tops the list. Yet what also matters is how these three tasks are carried out in practice—that is, how those responsible for the use of video surveillance really rank those two broad goals of securing evidence on the one hand, and preventing crime (points 2 and 3 both concern prevention) on the other. Preventing crime by using video surveillance could be done in two ways.

First, it is easy to envisage the mere presence of cameras as a crime deterrent. The key assumption here is that potential law breakers will be deterred from committing a crime simply because they will be aware that they may be watched.

Second, crime prevention could also be carried out in a more indirect way, by quickly reacting to unfolding events through continuous, real-time surveillance. This will not obviously achieve complete prevention, as the police will likely be reacting to crimes already underway or committed. Crime prevention is defined here as making sure that the criminals are caught in the act, arrested, identified and prosecuted, and in this way deterred from further criminal activity. The act of arresting and successfully prosecuting offenders could then be publicised as a success story to further strengthen the future deterrent effect. So the crime prevention effect of video surveillance here becomes essentially a matter of future hope.

There is arguably a significant difference between the two approaches to prevention, because they appear to put focus on different things. Moreover, it could be argued that these differences in focus may have their important justifications in different judgements about the trustworthiness of those subjected to video surveillance.

In the first case, video surveillance is supposed to deter people from committing crime by its mere presence, which would be a typical, simple case of situational crime prevention. The main focus here would be to make as much of a buzz about the presence of video surveillance as possible, mostly by posting information and warning signs. This method would be used in relatively limited areas, where committing some offence may be especially tempting—for example, in shopping malls or some sections of a motorway with drastic speed limits. In the first instance, its function would be to deter customers from shoplifting, and in the second to deter them from speeding. By its very nature, this surveillance philosophy could only work in areas of limited size. If the method were to be used too extensively, it would naturally lose its deterrent effect, as people would either stop paying attention to the information signs or the omnipresent video surveillance would lose its credibility. It could only work effectively by concentrating on certain 'hot spots'. Limiting the area of surveillance in this way may also reveal the underlying assumption about the trustworthiness of the individuals under surveillance; the leading assumption with such an approach appears to be that people do tend to behave correctly most of the time, and only in certain situations may they be tempted to act indecently. The surveillance measures are there to 'help them' stay on the right track. Deterrence of crime—not the prosecution of offenders—appears to be the prime purpose of the video cameras.

As for the second, more indirect approach to crime prevention by deploying video cameras, the focus would be more strongly on the prosecution of offenders. The surveillance would likely not be limited to certain, well-defined public areas, and would probably be relatively extensive, while the visible on-site information about the presence of video cameras might be scarce. The underlying assumption about the trustworthiness of individuals would be different too, as there would be widespread pessimism about their good nature. It could be assumed that these individuals were prone to committing offences and crimes. By subjecting them to monitoring, there would at least be a way to prosecute them, and in this way make them accountable for their actions. Once this was accomplished they would hopefully learn lessons and abstain from further destructive activity, because they would know that free-riding of any sort is unlikely to pay off due to the effective monitoring. In other words, there would be a substantial focus put on amassing evidence of crime for later prosecution, even though this would not necessarily have to be the stated primary goal of video surveillance.

In summary, I have here sketched two philosophies of public video surveillance with different sets of priorities and, more importantly, different assumptions about the trustworthiness of those under surveillance. If the distinction between the two philosophies is not sufficiently clear, it would be useful to look into the work of Garland and his distinction between the present culture of crime control and the 'penal welfarism' which characterised much thinking within Western criminal justice systems in the decades after World War II (Garland 2000, 2001; see also Crawford 2009). Whereas in the era of penal welfarism, where the problems of crime were viewed as socially created, and were to be fixed by means of social engineering and rehabilitation, the new philosophy takes a different stance that I will outline below.

First, the new approach places a much heavier emphasis on situational crime prevention, and assumes that basically everyone could potentially be a criminal under certain circumstances (Garland 2001). It is thus the task of public institutions, as well as institutions of civil society, to discourage such behaviour. Clearly visible video surveillance in 'sensitive' areas could fulfil such a role.

However, the new 'culture' of crime control also has another face, which is more pessimistic concerning the prospects for rehabilitation of convicted criminals. As a consequence, this new philosophy relies much more heavily on the prison system, not primarily as a means of rehabilitation but as a means of protecting the public from crime as well as a means of retribution. Demands for retribution are driven by politicians and find support in broad layers of the general public, which in turn is driven by the fear of crime and the will to give priority to the interests of crime victims rather than the perpetrators. The demand for retribution is a significant quality of this new philosophy, and a departure from the previous emphasis on rehabilitation.

There are thus two dimensions to the new 'culture' of crime control: the (re-)discovery of situational crime control on the one hand and a stronger emphasis on repressive measures against criminals on the other (which, by itself, can also be viewed as crime prevention measure). These two dimensions appear to correspond to the two approaches to video surveillance discussed earlier—that is, surveillance of limited 'sensitive' areas, where those subjected to surveillance are clearly informed about the measure, and a more widespread, diffuse video surveillance. Even though both approaches signal a feeling of distrust on the part of those who set up the surveillance systems, it could be argued that distrust appears more related to the second approach. The reason for this is that it embodies quite a pessimistic view of those subjected to surveillance, who are presumed to have *bad* intentions almost by default and need to be constantly monitored. Collecting evidence for a later prosecution is the primary goal of this kind of surveillance, not so much averting the crime by the mere presence of cameras.

Even though in the case of Poland it would be incorrect to talk about this being a 'new' philosophy of crime prevention (since it is unclear whether Poland, unlike its Western counterparts, has a prior history of penal welfarism), it is probably correct to say that this second approach has been informed by exactly this type of reasoning. That is, it has been informed by policies which view *punishment measures* as deterrents of future crimes, and finding and prosecuting law offenders becomes the primary goal. The criminal is fully responsible for his or her actions and therefore deserves isolation and, most of all, punishment. The retribution motive can be clearly seen as I review the way in which video surveillance techniques are being applied in Poland. The Polish, and some Western (for instance, British) approaches to crime prevention may bear similarities despite their divergent histories, which should hardly come as a surprise. If fear of crime is on the rise among the British public, it is after all a manifestation of diminishing trust, which could be the common denominator of the crime prevention strategies in both Poland and the UK.

Practical, real-life cases of video surveillance that would illustrate the type of approach to crime prevention just described would have the following characteristic: the video surveillance would be extensive and heavily tilted towards gathering evidence of crime rather than outright prevention and consequently there would be relatively little effort to inform the general public about the exact location of video the cameras.

The Polish setting appears to offer many examples of this kind of video surveillance, three of which will be reviewed here. The first involves municipal video surveillance in the city of Warsaw; the second, the use of video cameras to monitor forests; and the third, the use of video surveillance at an elementary school.

Case 3: Municipal Video Surveillance in Warsaw

Video surveillance in the Polish capital city of Warsaw is extensive, yet information about the presence of cameras is relatively scarce. Over seven hundred public surveillance cameras are located in different parts of the city, and the addition of more is being planned (Monitoring miejski 2012). The surveillance is being carried out by, among others, the Capital City Monitoring Centre (Stołeczne Centrum Monitoringu)—a company owned by the city of Warsaw. Its operators are located in fifteen Watch Centres, all of which except for one are located at local police stations (Gniadek 2009). The main task of the operators is to continually watch the camera feeds and, in case they notice anything suspicious, to alert the police. The images from the cameras, apart from being continuously watched by the operators, are stored safely for a limited period of time. In instances in which the recorded material could constitute evidence in a crime investigation, the storage time can be prolonged indefinitely. While the stated goal of the video surveillance system is *crime prevention*, the term itself is defined in a way that resembles the indirect prevention method described earlier. In other words, there seems to be little expectation of the crime-deterring effect of video equipment as such; rather, prevention is conceived of as the ability to react quickly to the events on the streets and to effectively prosecute the offenders. According to Jacek Gniadek, the manager of the Warsaw Monitoring Operation Centre, the system in Warsaw is superior to those in other cities like London or Berlin. The 'superiority' is to be found in the fact that the camera footage is being continuously watched and controlled by real people, whereas in many other cities the cameras scan their surroundings automatically. This 'passive' system is not very effective since, to be able to prosecute somebody for a crime, the perpetrator would have to look straight into the camera lens—something criminals do rather unwillingly, according to Gniadek. In Warsaw, on the other hand, it is possible to catch criminals in the act (Janiszewski 2010). Effectiveness in preventing crime is thus measured by the chance of identifying, arresting, and prosecuting the perpetrator. The system's ability to deter prospective criminals from committing a crime in the first place plays, at best, a secondary role.

Instances of catching offenders in the act are being used by the police to further boost public support for the surveillance system. In the spring of 2010, visitors to a police website could read a story about a group of youngsters who, on their way home from a party, decided to take off their clothes and walk naked on the streets of Warsaw (Waglowski 2008). In the darkness of night the youngsters could perhaps have done so unpunished but, thanks to the watchful eye of video surveillance, it was all recorded and noticed by a camera operator who alarmed the police. The youngsters were soon met by police officers and were ordered to get dressed. On the website there is a thorough description of the incident as a big success story for the municipal video surveillance, and the reader can learn how many offences and crimes are detected and recorded every year thanks to the watchful eyes of the cameras and their operators. In order to dispel any doubt as to what really happened, and what kind of offence had been detected and averted, the reader was even presented with the possibility of downloading high-resolution pictures of the naked persons—the ultimate material proof. It is also quite easy to read the underlying message: people tend to do silly things when they think that nobody is watching them, and now, perhaps, they have learned their lesson. The message is clearly enunciated in an interview with the camera operators, published in the Polish edition of *Newsweek* in March 2010. A camera operator put it quite frankly: "People can do really stupid things when they think nobody can see them" (quoted in Janiszewski 2010).

Case 4: Video Surveillance in a Forest

In Poland, as in most European countries, dumping garbage in forests is strictly prohibited. Since littering forests with rubbish, especially along roadways, has become an increasing problem, a local forestry administration body in one region (*voivodeship*) in central Poland decided to deploy a video camera. A short report about the measure appeared in the summer of 2009 in the local daily newspaper *Gazeta Pomorska* (Talaśka-Klich 2009). The camera was aimed at a parking space where motorists can relax during a journey and, according to a representative from the forestry administration, no law is broken because the visitors are clearly informed about the presence of the camera. All the actions of the visitors are recorded. A forestry administration official said that, even though 'sneak-peeking' was never the aim of the measure, the operators of the camera "can see a lot". On one occasion, they were able to see a woman who decided to defecate in the forest even though a wooden toilet was quite nearby—that record was deleted. However, no records of persons dumping rubbish into the forest are removed; instead, the forestry administration sends the litterers a fine with a photo. The size of the fine to be paid varies with the number of bags that have been dumped (100 złoty for five bags, 200 złoty for ten). Confronted with the photo evidence, most of the offenders promptly pay their fines. In one instance, a driver questioned the evidence but, having received an invitation to watch the video recording together with forestry administration officials, quickly changed his mind and paid the bill. Only one person has so far failed to pay the bill—and the particular reason for that was that the individual had died before the payment was due.

Even though the setting of the above case is different, one feature appears to be similar to the previous case: there is a deep conviction that the offenders were brought to justice thanks to the recorded evidence against them. There appears also to be a feeling of satisfaction that the forestry administration has finally got the right tools to crack down on the litterers. Outright prevention of the illegal dumping practices by the deterring presence of the cameras does not seem to be the major consideration behind the measure. The leading logic seems to be instead that people simply misbehave and will continue to do so until it becomes apparent to them that their actions are being monitored and that consequences will follow. The latter goal can only be achieved by careful evidence-gathering and harsh sanctions. Then, and only then, will the perpetrators hopefully learn their lessons and improve their behaviour.

Case 5: Video Surveillance at an Elementary School in the Town of Wąbrzeźno

Wąbrzeźno is a town located in northern Poland. One of the three elementary schools in the town had been equipped with video surveillance equipment thanks to a special grant from the education ministry.[2] The official aim of the grant and the main function of the video surveillance equipment were to stop unauthorised persons from entering the school, and thereby to stem any possible inflow of drugs. There were several cameras installed inside the building near the entrances, and one of the cameras was installed in the schoolyard. According to the headmistress of the school, drugs had not been a big issue for the school up to that time, so the video equipment was installed to address a possible future problem. When asked what the main benefits of the system have been so far, she mentions two. First, the school is finally able to do something about the vandalism of its property. This is not to say they have been able to prevent the devastation, which has in fact continued even after the deployment of the cameras. What is different now—and this appears to be a substantial difference to the headmistress—is that today the school is able to find those responsible and bring them to justice. To illustrate this, she recounts two such success stories; in one case they were able to trace those responsible for vandalising a school toilet, and in the second instance the cameras helped to find the youths who had covered one of the school's newly painted walls with graffiti (even though the perpetrators were masked!).

Another benefit of the system according to the headmistress is the possibility of supervising her co-workers. As a headmistress, she now has the right tools at her disposal to check whether the teachers appear on time in the classrooms, and whether or not they have forgotten about their duty to supervise pupils in the schoolyard during breaks.

Overall, the headmistress is very pleased with the surveillance system, adding that its only deficiency so far has been the lack of even more cameras. Also, this case then seems to fit the pattern: even though the original stated purpose of the video equipment was to better control access to the school, in practice the system's main role again appears to be that of collecting material evidence to be used against those who destroy school property. And, since the system has already proved useful in gathering such evidence against vandals, the headmistress is very satisfied with its performance. The damage being done to the school property has not been stopped, but the vandals can finally be tracked down and handed over to the police. A fringe benefit of the system—the possibility of supervising the teaching staff—may be a further confirmation of the culture of distrust, which may have created the demand for the surveillance system in the first place.

CONCLUDING DISCUSSION

The purpose of this chapter has been to put forward a particular explanatory model for the popularity of video surveillance. The model focuses on the low levels of interpersonal trust (low social capital) as a driver for the increasing and continued deployment of video surveillance in public space. I have argued for the plausibility of such a model by considering, on a theoretical level, the function of trust in the well-being of institutions, and the way in which a *lack* of trust can be compensated for in institutional settings. Since the lack of trust can partially be compensated for by increased monitoring, the use of video surveillance could well constitute a useful monitoring device. After this brief theoretical discussion I considered a number of possible indicators of interpersonal distrust as a major (if not primary) driving force behind the deployment of video surveillance cameras. I decided to look for these indicators in the stated, official purposes of surveillance, as well as in the actual practice of surveillance and the perceptions of surveillance goals by its operators. Considering the limited methodological arsenal I used, the conclusions at this point are necessarily at an early, interim stage. Nevertheless, the cases I have considered are of sufficient weight to warrant further investigation. The preliminary conclusion is that interpersonal distrust could indeed be a major driving force behind the increasing deployment of cameras in public spaces. To further verify and refine these findings more research is needed. Apart from reviewing more cases of video surveillance—for instance, in Poland or other comparable countries—and analysing the goals as well as the practice, it would be instructive to carry out a large-scale comparative study of the major democratic industrialised nations, in which quantitative data on trust and surveillance camera density (or public support for technical surveillance measures) would be checked for possible correlation. The cases presented in this chapter appear to confirm our theory-guided predictions concerning the balance between trust on the one hand and monitoring measures (as well as sanctions) on the other. Under circumstances of low levels of trust, monitoring measures gain prominence. Lack of trust creates demand for the second-best solution, which is the Hobbesian approach to the problem of free-riding that comes in the form of increased monitoring and threat of sanctions. As monitoring and sanctions are, at best, poor substitutes for trust, they will likely never be able to compensate for it, resulting in ever-increasing levels of monitoring and ever harsher threats of sanction. Under these circumstances, the popular demand for new cameras is likely to rise and be insatiable unless, of course, trust can be established. From this point of view, it would be interesting to follow the developments in Poland, and elsewhere, over an extended period of time, tracing a possible correlation between the popularity of video surveillance and levels of interpersonal trust. If a significant correlation between these two phenomena can be established, the popularity and density of camera surveillance in a country could perhaps be used as a measure of trust, giving us also valuable data on the cost of transacting in a particular society.

A final remark concerns the discussion on the effectiveness of video surveillance in preventing crime. The effects of video cameras on crime rates in areas where they are installed have been questioned (see Waszkiewicz's contribution, this volume, and also Waszkiewicz 2011). The questions posed in such studies are surely justified, particularly when crime prevention has been the stated goal of investing substantial amounts of taxpayers' money in the equipment. Still, one could argue that such critical studies may be missing a crucial point: crime deterrence may be one substantial goal of surveillance but perhaps equally significant is the detection of suspects, bringing them to justice, and—last but not least—exacting retribution. The latter goal appears to play a very substantial part in the eyes of camera operators in Warsaw and elsewhere. In this second role, the usefulness of cameras appears to be indisputable (Miłosz 2008). The fact that the camera operators and public see the usefulness of video surveillance primarily in terms of tools for tracking down offenders is hardly surprising in view of our earlier discussion about distrust in society and its increased need of monitoring. A distrustful society will be obsessed with cracking down on free-riders, and video cameras may offer it a welcome helping hand in pursuing that goal. Under such circumstances, the public will surely be willing to bear the high cost of surveillance, measured both in terms of taxpayers' money and a possible loss of privacy.

APPENDIX: INTERPERSONAL TRUST IN POLAND, SWEDEN, AND GERMANY. CUMULATIVE DATA FROM THE EUROPEAN SOCIAL SURVEY, 2006 AND 2008.

Of the three countries—Sweden, Germany, and Poland—Sweden appears to posses the largest reservoir of interpersonal trust. Even though there is a local 'peak' in the middle of the eleven-degree scale, indicating that a substantial number of respondents are neutral in their opinion, a clear majority

Table 5.1 Interpersonal Trust I

Country	Poland	Sweden	Germany
"You can't be too careful"	8.6	1.3	6.4
1	7.2	1.0	3.0
2	10.7	2.1	7.7
3	14.5	5.4	12.8
4	10.4	6.7	11.0
5	23.9	16.8	21.5
6	7.4	10.8	12.0
7	8.7	24.8	13.3
8	5.8	21.4	8.7
9	1.4	6.8	1.9
"People can be trusted"	1.5	2.8	1.8

The survey question read, "Would you say that most people can be trusted, or that you can't be too careful in dealing with people?" The valid percentage of respondents who chose the different alternatives.
Source: European Social Survey.

Table 5.2 Interpersonal Trust II

Country	Poland	Sweden	Germany
"Most people try to take advantage of me"	4.5	0.7	1.5
1	4.9	.6	1.3
2	7.7	2.1	3.8
3	11.7	3.6	7.3
4	9.9	4.9	8.8
5	26.3	14.9	22.8
6	9.4	11.7	12.6
7	10.9	24.6	18.2
8	9.5	24.0	16.0
9	3.0	9.0	4.5
"Most people try to be fair"	2.0	4.1	3.2

The survey question read, "Do you think that most people would try to take advantage of you if they got the chance, or would they try to be fair?" The valid percentage of respondents who chose the different alternatives.
Source: European Social Survey.

leans towards the upper end of the scale (the lower end of Tables 5.1 and 5.2)—the end which is meant to express trust. And so, a substantial majority of Swedes are willing to agree with the statements that "most people can be trusted" and that most people "try to be fair". The remaining two countries—Germany and Poland—appear to be far more distrustful than Sweden. In both countries, over 20% of the respondents take no stance, by choosing the middle point (5) on the scale. Of those who do take a stance, a majority of both Germans and Poles appear to disagree with the statement that "most people can be trusted". Here, however, we can also see a clear difference between these two countries; whereas in Germany slightly over 40% of respondents sympathise with the statement that you cannot be too careful in dealing with other people, the corresponding figure for Poland is over 50%. Table 5.2 reveals more differences. In Germany, about 22% of respondents tend to agree with the statement that most people would take advantage of other people; the corresponding Polish figure is over 38%. All in all, German respondents appeared to be more trustful than the Polish ones, even though the differences between these two groups of respondents were smaller than that between each of them and the Swedish group. Of the three reviewed countries, Poland appeared to be the least trustful one.

NOTES

1. The term *tragedy of the commons* is often attributed to Hardin (1968).
2. The description of this case is based on the interview with the headmistress of Elementary School Number 3 in Wąbrzeźno, Ewa Dorau (Dorau 2009).

REFERENCES

Bibliography

Arrow, Kenneth. 1974. *The Limits of Organization*. New York: Norton.
Coleman, James. 1990. *Foundations of Social Theory*. Cambridge, Mass.: Harvard University Press.
Crawford, Adam. 2009. "Situating Crime Prevention Policies in Comparative Perspective: Policy Travels, Transfer and Translation." In *Crime Prevention Policies in Comparative Perspective*, ed. Adam Crawford. Portland, Ore.: Willan, 1–37.
Garland, David. 2000. "Ideas, Institutions and Situational Crime Prevention." *In Ethical and Social Perspectives on Situational Crime Prevention*, ed. Andrew von Kirsch, David Garland, and Alison Wakefield. Portland, Ore.: Hart, 1–14.
———. 2001. *The Culture of Control*. Chicago: University of Chicago Press.
Garton Ash, Timothy. *The Polish Revolution: Solidarity*. London, England: Penguin.
Hardin, Garret. 1968. "The Tragedy of the Commons." *Science* 162(3859): 1243–48.
March, James, and Johan Olsen. 2004. *The Logic of Appropriateness*. Arena Working Papers WP 04. Oslo: ARENA Centre for European Studies.

Misztal, Barbara. 1996. *Trust in Modern Societies*. Cambridge: Polity Press.
Neyland, Daniel. 2006. *Privacy, Surveillance and Public Trust*. Houndmills Basingstoke Hampshire: Palgrave Macmillan, 2006.
Ostrom, Elinor. 1990 *Governing the Commons: The Evolution of Institutions for Collective Action*. New York: Cambridge University Press, 1990.
Putnam, Robert. 1996. *Making Democracy Work: Civic Traditions in Modern Italy*. Princeton, N.J.: Princeton University Press.
———. 2000. *Bowling Alone: The Collapse and Revival of American Community*. New York: Simon and Schuster.
Rothstein, Bo. 2008. "Tillit och socialt kapital." In *Jämförande politik*, ed. Jessica Lindvert. 218–243: Liber, Malmö.
Sztompka, Piotr. 1999. *Trust: A Sociological Theory*. Cambridge: Cambridge University Press.
Talaśka-Klich, Lucyna. "Wielki Brat czai się między drzewami." *Gazeta Pomorska*, 24 June.
Waszkiewicz, Pawel. 2011. *Wielki Brat Rok 2010*. Warsaw: Oficyna.

Electronic Sources

Egzamin na prawo jazdy z kamera i mikrofonem. 2009. "Egzamin na prawo jazdy z kamera i mikrofonem. Wirtualny Nowy Przemysł—Wiadomości." *wnp.pl. Portal Gospodarczy*, May 9. Available at http://www.wnp.pl/wiadomosci/9107.html.
Internetowy System Aktów Prawnych. 2012. "Dz.U. 2009 nr 97 poz. 803." [Polish Parliament Website]. Available at http://isap.sejm.gov.pl/DetailsServlet?id=WDU20090970803.
Janiszewski, Bartosz. 2010. "Warszawiaku, oni cię znają z widzenia," *Newsweek.pl*, 21 March. Available at http://www.newsweek.pl/artykuly/sekcje/spoleczenstwo/warszawiaku-oni-cie-znaja-z-widzenia,55518,1.html.
"Łowcy Skór." 2012. "Łowcy Skór." *Gazeta Wyborcza*, 1 March. Available at http://wyborcza.pl/0,82452.html.
Miłosz, Maciej. "Monitoring nie zwiększa bezpieczeństwa?—Onet Wiadomości." *Onet.pl*, 6 November. Available at http://wiadomosci.onet.pl/kiosk/kraj/monitoring-nie-zwieksza-bezpieczenstwa,1,3348945,wiadomosc.html.
Pochrzęst, Agnieszka. 2006 "Z Kamerą Na Chorego w Karetce" *Wirtualna Warszawa*, 22 September. Available at http://www.wirtualna.warszawa.pl/w/060922/z-kamer%C4%85-na-chorego-w-karetce.
Pochrzęst, Agnieszka. 2007 "Kamery Już Nie Nagrywają Chorych" *Gazeta Wyborcza*, April 5. Available at http://warszawa.gazeta.pl/warszawa/1,86775,4042640.html.
Monitoring miejski. 2011. "Monitoring miejski—Bezpieczna Warszawa," Bezpieczna Warszawa [Warsaw Mayor's Office Website]. Available at http://bezpieczna.um.warszawa.pl/bezpie publiczne/monitoring-miejski.
Waglowski, Piotr. 2008 "W jaki sposób zdjecia z monitoringu trafiły do mediów?" *vagla.pl*, 27 May. Available at http://prawo.vagla.pl/node/7893.

Interviews

Dorau, Ewa. 2009. Interview with Ewa Dorau, headmistress of Elementary School Number 3 in Wąbrzeźno, Poland, April 2009. Written summary in Swedish available at Södertörn University, Flemingsberg, Sweden.
Gniadek, Jacek. 2009. Interview with Jacek Gniadek, manager of the Warsaw Monitoring Operation Center," 17 December 2009. Written summary in Swedish available at Södertörn University, Flemingsberg, Sweden.

6 How Effective is the Public Video Surveillance System in Warsaw?

Paweł Waszkiewicz

INTRODUCTION

Fear of crime (becoming a victim of any kind of criminal behaviour), which nowadays may be perceived as an old phenomenon tightly associated with living in society, became a concern of public opinion less than fifty years ago in the United States (in the 1960s) and forty years ago (in the 1970s) in the UK (Newburn 2007: 355). However, some scholars consider it as being present in Western culture from 'time immemorial' (along with fear of the streets and juveniles; Pearson 1983: 236). In Central and Eastern Europe a rising fear of crime was accompanied by political changes transpiring at the beginning of the 1990s. Between the first Polish survey on fear of crime (1987) and the first one conducted after the democratisation process of 1989 (1993) the number of people perceiving Poland as a safe country dropped from 74% to 26%. At the same time, the number of those who regarded their neighbourhood as unsafe increased from 17% to 30% (even in as late as 1993, as high as 67% of respondents felt safe in their place of residence; Feliksiak 2011: 1–2). Fear of crime reflects crime rates only to a certain extent; very often it "lives a life of its own" stimulated not only by popular media but also by economic factors. Nevertheless, it influences assorted life decisions based on assumptions made about crime and possible victimisation. Fear of crime also creates a 'demand' for safety, which is being addressed, on the one hand, by the private sector and, on the other hand, by the public sector. Politicians are trying to tackle those fears and expectations by promising to 'be tough on crime', a slogan that seems to be one of the most appealing to an electorate that feels increasingly unsafe. The prime aim of the 'law and order' approach is to reinforce criminal justice and to introduce longer mandatory sentences, including capital punishment. According to this paradigm, crime prevention is being limited to making potential offenders afraid of the outcome of their criminal actions almost to the same extent as of being punished. This approach reduces the significance of social control and increases the role of formal, technical means of control, which are supposed to deter the 'bad guys' from breaking the law (and order). Even with this limited understanding of crime prevention, which does not confront the real causes of crime, there exists a chance of preventing certain crimes from being committed.

Crime prevention is defined in a number of ways, which may be classified in two groups of approaches (Newburn 2007: 566). The first, sometimes referred to as a medical model, distinguishes primary, secondary, and tertiary activities. Such typology concentrates on the targets and timing of the undertaken actions. In primary prevention they will involve the general public and activities commenced before the crimes are perpetrated, while secondary prevention focuses on high-risk groups and places (as a response to events that have already occurred); tertiary prevention concentrates on identified offenders and victims and should prevent them from recidivism and re-victimisation (Heinz 1998: 24). The second typology differentiates between situational and social approaches. Regardless of which crime prevention 'school' is chosen there exists a certain intersection—the goal shared by almost everyone is to prevent crime, a task usually understood as reducing the number of all crimes in a specified area or merely selecting types of offences (e.g., burglaries). Planning, decision making, and choosing methods and their implementation result from local, cultural, and political circumstances. If there exists a set goal (a reduction of crime rates) then there is a chance to measure the outcome of the undertakings. Effectiveness may be determined upon the basis of the extent to which goals have been achieved, making it possible to answer in particular settings the simple question, are crime prevention efforts effective or not? Such a procedure seems to be clear and obvious, but it is followed very rarely. In the majority of cases there is a lack of interest and willingness to be informed about the actual results. This approach could be described as a security paradox—security has become so important that almost no one is interested in the real evaluation effects on the one hand while, on the other hand, in our 'insecure and liquid times' many people perceive themselves as security experts. In 1996 the US Congress required the attorney general to provide a comprehensive assessment of the effectiveness of Department of Justice grants, which were to assist law enforcement agencies in crime prevention. The objective was for research to be "independent in nature" and to "use rigorous and scientifically recognized standards and methodologies" (Sherman 1998b: 3). The first report (which evaluated five hundred crime prevention projects) concluded, "The effectiveness of most crime prevention strategies will remain unknown until the nation invests more in evaluating them", adding, "By scientific standards, there are very few 'programs of proven effect'" (Sherman 1998a: 681).

RESEARCH ON VIDEO SURVEILLANCE EFFECTIVENESS

To some extent not much has changed since 1996 concerning worldwide evaluation of the effectiveness of different crime prevention activities. The same problem is faced by research on video surveillance effectiveness conceived as a crime deterrent. The popular belief in public video surveillance as a 'silver bullet' that will 'fight crime' is being 'taken for granted'. In Britain (labelled as a video surveillance 'capital') politicians and practitioners were so convinced of the cameras' efficiency that when at the beginning of the 1990s the state was engaged in funding the mass expansion of public video surveillance there were no systematic assessments of its prevention and detection effectiveness (Norris 2010: 401). The first systematic review of the crime prevention effects of video surveillance was carried out in 2003 by Welsh and Farrington, whose meta-analysis included only evaluations meeting the following criteria:

- Video surveillance was the focus of intervention.
- There was an outcome measure of crime.
- The evaluation design was of high methodological quality, with the minimum design involving before-and-after measures of crime in experimental and control areas.
- There was at least one experimental area and one comparable control area.
- The total number of crimes in each area before intervention was at least twenty. (Welsh and Farrington 2002: v)

Half of the twenty-two evaluations showed a desirable effect on crime (an achieved significant decrease in crime), five demonstrated an undesirable effect on crime (a considerable increase of crime), five disclosed no effect on crime, and one was classified as finding an uncertain effect on crime. The average effect (from nine projects implemented in city centres or public housing) was a statistically insignificant odds ratio of 1.02 (Welsh and Farrington 2002: 27).

The same quasi-experimental approach was adopted in the evaluation conducted by Gill and Sprigs in their study of fourteen public video surveillance implementations across Great Britain. These findings do not support video surveillance crime prevention effectiveness, and conclude, "The CCTV [closed circuit television] schemes that have been assessed had little overall effect on crime levels. Even where changes have been noted, with the exception of those relating to car parks, very few are larger than could have been due to chance alone and all could in fact represent either a chance in variation or confounding factors" (Gill and Spriggs 2005: 43). No link was established between video surveillance and the reduction of fear of crime. An interesting finding asserted that respondents "who were aware of the cameras admitted higher levels of fear of crime than those who were unaware of them" (Gill and Spriggs 2005: 60). The Home Office–sponsored report claims, "CCTV is an ineffective tool if the aim is to reduce overall crime rates and make people feel safer. The CCTV systems installed in 14 areas mostly failed to reduce crime (with a single exception), mostly failed to allay public fear of crime (with three exceptions) and the vast majority of specific aims set for the various CCTV schemes were not achieved" (Gill and Spriggs 2005: 61).

Certain studies attempting to evaluate the impact of public video surveillance on crime and the feeling of security adopt the same quasi-experimental schemes but deal with a different audience. Research conducted in Malaga, Spain, intended to measure public video surveillance in the city's Old Town section, but in the experimental and control areas it examined not the residents but casual passers-by (Miller 2007: 1–4). Undoubtedly, it is easier to conduct such research, but it seems that random people who may appear on the site (tourists, especially considering that this is Old Town) cannot be perceived as the most authoritative source of information. In addition, the use of such a model may cause a large sample fluctuation—for example, during the first phase (before installation) the number of tourists could be higher than average, and lower than average during the second phase. The weather can produce the presence in the street of representatives of certain groups and the absence of others (non-pedestrians being under-represented, while a group of workers will be over-represented in the vicinity of the shops and offices employing them). Researchers in Malaga gathered data from four different sources: police statistics and a victimisation survey, as well as interviewing business owners and operators of the public video surveillance system. Police statistics and victimisation data were gathered from four research areas, each consisting of ten streets. Twenty streets were defined as an experimental area—all in the Old Town section. In half of those experimental streets public video surveillance cameras were installed a year after the first phase of the survey, and ten streets were adjacent to the first group, but no cameras were installed. Ten control streets were comparable to ten streets with cameras, but no cameras were installed in those locations and ten control streets were comparable to ten experimental streets without cameras. Such a structure of the survey may measure differences in the geography of crime—that is, the spatial displacement of crime. According to police statistics, the number of registered offences in the experimental area with cameras decreased from 982 to 963 (1.9%); at the same time, the number of crimes in the experimental streets without cameras increased from 755 to 852 (14.6%). In control streets comparable to streets with cameras, the number of crimes increased from 560 to 622 (11.1%), and in comparable control streets without cameras it fell from 751 to 740 (1.4%; see Cerezo Dominguez and Diez Ripolles 2010). Taking into account their limitations these results do not support the thesis on the crime-deterrent influence of cameras. The registered change was statistically insignificant and, at the same time, there was a comparable decrease of crime in the ten control streets without cameras. The only assumption, which the results in question may confirm, is the possible displacement of certain offences from those streets where cameras are installed into adjacent ones without cameras.

Concentrating on passers-by instead of residents does not exclude the validity of research on security perception among casual passers-by—for instance, the one carried out in Reeperbahn, Hamburg (Zurawski 2007: 52). This scheme cannot capture changes caused by certain forms of intervention, such as installing cameras, and is not intended to achieve such a result; nonetheless, it may answer questions about the perception of the effectiveness of public video surveillance systems in specific settings. It seems very interesting that at a time when more than two-thirds of respondents (66.4%) support general video surveillance only 43% feel that cameras would protect them against crime (Zurawski 2010: 269). The biggest support for public video surveillance was shown by those respondents who declared feeling safe in the given area (Zurawski 2010: 267). The author summarised these findings as follows: "Results from a survey conducted in Hamburg in 2006 suggest that closed-circuit television (CCTV) has little to do with manufacturing security/feelings of safety among people", and concluded, "The expectations attached to CCTV often do not correspond to the roots of feelings of insecurity" (Zurawski 2010: 259, 274).

In 2009 Rothman, in his research dealing with Vienna, adopted a methodology similar to that of the Hamburg survey. He selected two squares where public video surveillance cameras had been installed and then conducted interviews with passers-by older than sixteen years of age ($n = 317$; Rothman 2010: 104). The respondents were asked, for example, how safe they feel at night and how they estimate the chances of becoming victimised in their area of residence. The results were compared to knowledge about the public video surveillance system applied in the surveyed area. Paradoxical results were obtained in one of the squares (Schwedenplatz); knowledge about a functioning video surveillance system was correlated with a significantly higher feeling of insecurity: 82% respondents among those who knew about the video surveillance system felt secure, as did 92% of those who were unaware of the system's existence. At the same time, the perception of respondents in the second square (Karlsplatz) was the reverse: those who knew about the video surveillance cameras felt more secure (77%) than those who were unaware of the cameras' existence (73%; Rothman 2010: 105).

Even taking into account the limitations of the summarised surveys, none support the 'common knowledge' that the public video surveillance system is effective in reducing the number of crimes committed in the area of operation or increasing the perceived feeling of safety.

RESEARCH AREAS IN WARSAW—
SELECTION PROCESS AND DESCRIPTION

The choice of methods and research areas is crucial and strongly influences the results of each study—and by no means exclusively those on crime prevention research. Therefore, not only was the research tool prepared with great care—the choice of the research areas was preceded by the development of selection criteria. The main criteria were:

- The complex nature of the city area, not occupied only by apartment buildings but also including commercial and entertainment facilities.
- The average level of security, which differs within city borders (choosing either very safe neighbourhoods or 'ghetto-like' ones will result in the gathering of social artefacts).
- No past presence of public video surveillance cameras in the area.
- Public video surveillance as the focus of intervention.
- A possible comparable area (concerning architecture, demography, and economy) without public video surveillance cameras.

All these criteria were met in certain neighbourhoods in Wola, one of seventeen Warsaw districts. This is a highly urbanised area with the characteristic features of a modern city: residences, commerce, services, and entertainment. It is also adjacent to the Warsaw city centre—the Śródmieście district. The adoption of a quasi-experimental design—that is, as used in studies evaluated by Welsh and Farrington (2002) and Gill and Spriggs (2005)—requires a selection of comparable experimental and control areas. Experimental areas in this case were those in which a variable (the video surveillance camera), whose impact was to be examined, was introduced. Control areas were devoid of the variable, but comparable in other respects: security, urban planning/architecture, economy, and demography.

Two experimental areas and two control areas (one for each district) were selected after an analysis and a discussion held within the Department of Criminalistics at the University of Warsaw. The survey financing was limited, and at most four areas could be analysed. This should have minimised risks associated with conducting a comparison of results from only one experimental and one control area—such an outcome could be random. The first pair of areas bordered on the Warsaw Central Railway Station (Warszawa Centralna). The second was chosen from a more residential part of town—the Muranów district. The centre of the first experimental area (E1) was the crossroads of Żelazna Street and Chmielna Street, where a public video surveillance camera was installed a month after completing the first phase of research. All areas were in the shape of a circle with a radius of 150 meters. The centre of the control area (K1) for E1 was the intersection of Sienna Street and Miedziana Street. The centre of the second experimental area (E2) was the crossroads of Anielewicza Street and Smocza Street, and its control area (K2) centre was the intersection of Nowolipki Street and Smocza Street. Each area included a random sample of adult residents ($n > 100$) filling in anonymous questionnaires distributed and collected door-to-door. One of the consequences of the questionnaires' anonymity was that certain respondents were questioned only once (during one phase of research). The questionnaire consisted of twelve questions, of which two were changed between the survey phases. Questions about self- victimisation and the victimisation of children and the feeling of security constituted the main part of the study. Their task was to measure the actual impact of installing cameras on crime (reported victimisation) and the feeling of safety.

PUBLIC VIDEO SURVEILLANCE IN WARSAW

The Warsaw public video surveillance system is definitely the biggest in Poland and comparable to the largest in Continental Europe. Its predecessor was the video surveillance system installed at the end of the 1970s at the Central Railway Station. The present-day system differs in many respects: size, purpose, and technical abilities. A system observing public space in all seventeen Warsaw districts was introduced in 2001. Since then it has been financed, administered, and operated by the City of Warsaw's Crises Management and Safety Department, with full access for the Warsaw Metropolitan Police. About eight hundred city cameras allow their operators to monitor and record all actions undertaken in their surroundings. (By way of comparison, the system in Cracow, a city of 756,000 inhabitants, consists of seventeen cameras, while in Toronto, with a poulation comparable to Warsaw—2.6 million in Toronto versus 2 million in Warsaw—there are twenty-three police cameras). It is worth mentioning that all cameras are monitored live '24/7', which should make them more effective. The cameras in question monitor only public space (though there are thousands of additional cameras inside city buildings, buses, and trams, not to mention private estates). The annual cost of maintaining such a public video surveillance system (with more than two hundred operators) exceeds €4.5 million. The prime aim of the system is crime prevention. Although the city covers all the costs of installing the system and bears the on-going expenses, until 2006 there were no empirical studies determining whether the goal had been achieved. This did not stop the Crises Management and Safety Department from making the following statement on its website: "There is no doubt that public video surveillance is effective in fighting crime. According to the Warsaw Metropolitan Police, crime in areas covered by public video surveillance has dropped after installing the cameras by 50–60%" (Crises Management and Safety Department, City of Warsaw 2012). Information obtained from the Warsaw Metropolitan Police claims that public surveillance video effectiveness has not been evaluated. This lack of assessment and UK findings were the key factors in deciding to commence such research in Warsaw in 2006.

Victimisation

It should be noted that the police do not posses specific data on crime for the surveyed areas—only data for whole districts or bigger neighbourhoods were available. This was one of the reasons why the analysis solely encompassed data obtained from the residents. It is also possible that a paradoxical increase in the number of crimes recorded by the police and generated by efficiently working camera operators actually reduces the dark figure of crime. In the case of an anonymous victimisation survey we are not compelled to deal with crimes unreported by the victims. The only category that is not fully included in such data is that of public disturbances (e.g., alcohol consumption in public or vandalizing public utility buildings), since the respondents may not perceive themselves as direct victims.

It was assumed that the anticipated change would take place in the experimental area if public video surveillance proved to be effective in reducing crime and increasing the feeling of security. Both changes should transpire to a greater degree in the experimental areas than in the control ones. Another case to be considered in favour of cameras envisaged as a preventive measure is the growing number of crimes and the decline of confidence in the experimental areas, albeit to a lesser extent than in the control areas. In order to avoid an over-interpretation of the results only the percentage change was compared. Meta-analysis was adopted in the manner suggested by Farrington and Welsh—that is, the number of crimes (and offences) in the experimental and control areas reported prior to public video surveillance intervention was compared to the number reported subsequently.

The survey on respondents, the victimisation of their children, and how safe and secure they feel was based on answers to six questions (the same during both phases of research). The first question concerned the victimisation experiences of the respondents in the course of twelve months before the survey (in both phases). The number of reported offences and the calculated rate were compared (the size of the samples varied slightly, with a minimum of one hundred respondents, which causes the uneven weight of individual responses) in order to show the changes accurately. The value obtained from the first phase was considered as 100%.

The residents of all studied areas were significantly less victimised in 2007 than in 2006 (this simplification is legitimate taking into account the fact that both phases of the study were conducted during the later months of the calendar year—October and November). The decrease in the number of residents who became victims of crimes or offences ranged from 37.7% (area E2) to 71.4% (area K2). As can be seen in Table 6.1, the situation of the control area residents greatly improved as compared to those living in the experimental area.

The presented results are insufficient, which may cause their over-interpretation in terms of measuring public video surveillance effectiveness. The respondents were victimised not only in the areas of research, but also in other parts of Warsaw. Examining the impact of the cameras on security had to be taken into account. The results were analysed using a precise identification of the places (made by the respondents) where the offences had been committed. Results including only offences perpetrated within the areas of research are presented in Table 6.2.

Table 6.1 Resident Victimisation Rates in Warsaw During 12 Months Preceeding the Survey

Area	Phase 1 (2006)	Phase 2 (2007)	Percentage change between Phase 1 and 2
E1	11.7%	5.9%	-49.6%
K1	15.1%	9%	-40.4%
E2	17%	10.6%	-37.7%
K2	21%	5.8%	-71.4%

Note: Sample size (n) E1=102 (Phase 1), 100 (Phase 2); K1=105 (Phase 1), 100 (Phase 2); E2=100 (Phase 1), 111 (Phase 2); K2=117 (Phase 1), 102 (Phase 2).

Similar to victimisation rates from the whole of Warsaw, the number of incidents (and percentage) of reported victimisation in each of the four areas also decreased. However, it is difficult to determine a relationship between the installation of cameras and this fact, since in both areas (experimental and control) the number of offences declined more in the area where cameras had not been installed. Obviously, dealing with such small numbers can cause accusations of low diagnostic-value evaluation, but it should be noted that the difference is perceptible in the number of crimes and expressed as a percentage.

The factor calculated for the meta-analysis (the ratio of reported offences before and after the installation of the cameras) is 1.33 for E1, 2.0 for K1, 2.5 for E2, and 3.75 for K2. The construction of this factor renders even more discernible that which was demonstrated in a comparison of the percentage change. We cannot conclude upon the basis of these data that cameras affected the actual safety of residents of those areas where they had been installed. On the other hand, we are not entitled to deduce that the installation of cameras caused a smaller decrease in the number of crimes in the experimental areas over the control areas. It appears that public video surveillance cameras did not achieve the expected results.

Table 6.2 Resident Victimisation Rates in the Research Area During 12 Months Preceeding the Survey

Area	Phase 1 (2006)	Phase 2 (2007)	Relative change
E1	3.9% (4)	3% (3)	-23.1% (−1)
K1	7.6% (8)	4% (4)	-47.4% (−4)
E2	10% (10)	3.5% (4)	-65% (−6)
K2	12.6% (15)	3.9% (4)	-69.1% (−11)

Note: Absolute numbers in brackets.

The respondents were exclusively the adult residents of the selected areas, although they were also asked about their children's victimisation. Presumably, parents are not aware of all instances in which their children become victims of crimes or offences, but are certainly informed about those violent acts whose effects cannot be easily concealed (battery, theft, or extortion).

Young people (especially adolescents and young adults) suffer from acts of violence significantly more often than older people. As Siemaszko notes, "A specified victimisation risk group is the one aged 16–19. It turns out that not only the criminals—as is known—but also their victims are recruited primarily from the younger age groups. However, it is worth mentioning that the regression analysis did not reveal a linear relationship to age" (Siemaszko 2001: 137). The calculation of the percentage of children whose parents were aware of their victimisation took into account only those respondents who declared that they have children; thus, the weight of absolute values (the number of crimes) is transformed into a presented interest in a relationship other than the victimisation of adult residents.

The children victimization results, as in the adult group, exemplify a change in (almost) all areas of research; drawing conclusions from them is unreliable due to lower absolute values. Reducing the number of reported crimes from two to zero (E2) may mean a decrease of 100%, while at the same time in another area (K2) a decline of six crimes (from eight to two) means a 75% decrease. This type of restriction arising from an analysis of a small number of crimes (in absolute terms) caused Welsh and Farrington to limit their evaluation to a program in which a minimum of twenty crimes was recorded prior to the introduction of the variable. While adult victimisation data are close to this value, making possible a more accurate analysis, only subsidiary data related to children may be used owing to methodological integrity.

In addition to the already mentioned limitations in the interpretation of results obtained from the small number of offences reported by respondent parents, mention should be made of one more restriction. As in the case of the victimisation of adult respondents, the limitation in question involves the place where the reported crime took place. Extracting crimes only from research areas produces even lower values, while a reduction of the number of crimes to a single one is equal to a 100% decline. This could account for the 'total elimination of crime' afflicting minors in the three research areas; naturally, such an approach is unauthorized.

As expected, the offences of which the parents (respondents) were aware were dominated by robbery, theft, and extortion, accompanied by the seizure of mobile phones and wallets as well as assault; the effects of such offences cannot be concealed from the parents.

The avoidance of conclusions does not restrict proposing a solution for the preparation of this type of research in the future. Two solutions would make it possible to obtain data for statistical analysis: the choice of a much larger sample, which will provide statistically significant material, or references also to minors. The first solution greatly increases the cost of research, while its significant limitation in terms of acquiring knowledge about only major crimes continues to be present. The second appears to be methodologically appropriate on the assumption that anonymity will be guaranteed to respondent children- (and also in relation to their parents) while obtaining consent from the parents for their offspring to participate in the research.

Feeling Secure

Feeling secure is a subjective appraisal of reality carried out by each person in terms of perceived threats. Assuming that public video surveillance cameras have a deterrent effect on crime, changes resulting from their installation should be reflected not only in the number of committed offences but also in the safety assessment carried out by the residents of the areas where they had been installed. Residents experiencing a lower crime risk should feel safer. Another potential and expected effect is a higher feeling of security resulting from an awareness of the use of new measures to combat crime on the part of relevant law enforcement agencies. Therefore, such feelings may have the nature of the placebo effect or result solely from a conviction about the effectiveness of certain measures (in this case, public video surveillance cameras) or a real, although subjective, assessment of safety.

Respondents were asked to identify how safe they feel in three different areas—Warsaw overall, the district/borough, and the yard/block—by selecting one of four answers: very safe, safe, unsafe, and very unsafe. There is a tendency to choose the median value, and thus the range of possible responses was constructed to prevent such choices by presenting an even number of possible answers. Respondents were 'forced' to decide whether they feel safe or unsafe. The indicated area includes a district or a settlement as part of the city, but for the purposes of this study it was used to test the well-known sense of security dependence; security in a wider area (the whole country, province, or city) is rated well below the region nearest to the respondent. Specific questions about different areas make it possible to compare changes in feeling safe at a national or regional level and within selected small areas.

Changes measured by questions about victimisation (a decreasing number of offences) were accompanied by an increase in feeling safe on the part of residents in all four areas (see Table 6.3). The percentage of people who felt safe and very safe in Warsaw grew. Because the sample sizes varied slightly between the two phases, the change in perceptions was calculated using relative values. Thus, the relative values obtained from the first phase were used as index for comparison with the second phase (example: respondents feeling safe or very safe in E1, 79%—the value of the Phase II study—represents a 13% increase relative to the per cent value obtained during

Table 6.3 Residents of Four Areas within Warsaw Borders Feeling Safe in 2006 and 2007

Area	Phase 1 (2006)		Phase 2 (2007)		Change (relative increase in safe and very safe)
	Very safe and safe	Unsafe and very unsafe	Very safe and safe	Unsafe and very unsafe	
E1	69.6%	30.4 %	79%	21%	+ 13%
K1	50.5%	49.5%	70%	30%	+ 40%
E2	62%	38%	74.8%	25.2%	+ 20.6%
K2	70.9%	29.1%	83.3%	16.7%	+ 17.5%

Note: Sample size (n) E1=102 (Phase 1), 100 (Phase 2); K1=105 (Phase 1), 100 (Phase 2); E2=100 (Phase 1), 111 (Phase 2); K2=117 (Phase 1), 102 (Phase 2).

Phase I, which in this case was 69.6%). The same procedure was adopted to calculate a shift in feeling safe in a given area/district or yard/block, and after nightfall (see below).

Fluctuations in feeling safe in each area were statistically significant and ranged from 13% (E1) to 40% (K1). Considering that in experimental areas (including E1) intervention took place in the form of a public video surveillance camera installation, it is difficult to observe its effects upon the basis of obtained data. This could be due to the fact that the question dealt specifically with feeling safe in Warsaw. If this is the reason, and the public video surveillance system influences the assessment of security in the vicinity of the intervention, changes should be observable in responses to subsequent questions.

The change in the perception of security was in (almost) every area similar to feeling safe in the city, statistically significant and ranging from 2.3% (K2; the only area with little change) to 26.4% (E1). This is one of the two examples at the time of the survey when the expected trend was observed in the chosen area. The change (increase) in feeling safe among the respondents in areas where public video surveillance cameras were installed was proportionately greater than in the control areas, where this variable was not introduced. Only while considering the second pair of areas is this difference significant—an increase in the group declaring feeling fairly safe and safe by 25.8% (E2) in relation to an increase in this group by 2.3% in K2. The difference between changes in areas E1 and K1 is very small—from 26.4% to 24.5%—but, as has been mentioned, this is one of two examples when the expected trend occurs in all areas (see Table 6.4).

The most important for the evaluation of the effectiveness of public video surveillance cameras is the change in the perception of safety in areas closest to the respondents: their backyard and block (see Table 6.5). Cameras were installed in the immediate vicinity, and their preventive impact

Table 6.4 Residents of Four Areas Feeling Safe in Their District in 2006 and 2007

Area	Phase 1 (2006) Very safe and safe	Phase 1 (2006) Unsafe and very unsafe	Phase 2 (2007) Very safe and safe	Phase 2 (2007) Unsafe and very unsafe	Change (relative increase in safe and very safe)
E1	68.9%	30.1%	87.1%	12.3%	+ 26.4%
K1	60.9%	39.1%	75.8%	24.2%	+ 24.5%
E2	71.7%	28.3%	90.2%	9.8%	+ 25.8%
K2	77.8%	22.2%	79.6%	20.4%	+ 2.3%

Note: Sample size (n) E1=101 (Phase 1), 101 (Phase 2); K1=105 (Phase 1), 99 (Phase 2); E2=99 (Phase 1), 112 (Phase 2); K2=117 (Phase 1), 103 (Phase 2).

should be greatest precisely in the field which they cover. This has not taken place. Significant changes were recorded as regards the sense of safety in three of the four areas, and the same holds true for the estate/area; the expected trend is discernible (in the experimental areas the sense of security improved proportionately more than in the control areas, although in one pair—E1 and K1—the difference was only 0.3%). However, these changes are smaller than those in data obtained for feeling safe in a district or the whole city. This may be due to the fact that the output values (measured during Phase I) were significantly higher for the immediate neighbourhood than for other defined areas. In this case, the increased number of people who felt safe, with the same absolute value in the same large sample, does not translate into an identical percentage change. If in both pairs underwent a change identical as in the second pair (E2 and K2) then we may conclude that an apparent difference occurred in the experimental and control

Table 6.5 Residents of Four Areas Feeling Safe in Their Yard and Block in 2006 and 2007

Area	Phase 1 (2006) Very safe and safe	Phase 1 (2006) Unsafe and very unsafe	Phase 2 (2007) Very safe and safe	Phase 2 (2007) Unsafe and very unsafe	Change (increase safe and very safe)
E1	83.3%	16.7%	91%	9%	+ 9.2%
K1	79.4%	20.6%	86.5%	13.5%	+ 8.9%
E2	81.6%	18.4%	92.9%	7.1%	+ 13.8%
K2	82%	18%	79.4%	20.6%	- 3%

Note: Sample size (n) E1=102 (Phase 1), 100 (Phase 2); K1=102 (Phase 1), 96 (Phase 2); E2=98 (Phase 1), 113 (Phase 2); K2=117 (Phase 1), 102 (Phase 2).

Table 6.6 Residents of Four Areas Feeling Safe from Dusk until Dawn in 2006 and 2007

Area	Phase 1 (2006)		Phase 2 (2007)		Change (relative increase safe and very safe)
	Very safe and safe	Unsafe and very unsafe	Very safe and safe	Unsafe and very unsafe	
E1	44.5%	55.5%	62.4%	37.6%	+ 40.2%
K1	34%	66%	51.5%	48.5%	+ 51.5%
E2	49%	51%	67.3%	22.7%	+ 37.3%
K2	65.3%	34.7%	68%	32%	+ 4.1%

Note: Sample size (*n*) E1=101 (Phase 1), 101 (Phase 2); K1=106 (Phase 1), 99 (Phase 2); E2=98 (Phase 1), 113 (Phase 2); K2=118 (Phase 1), 103 (Phase 2).

areas. Taking into account the slight difference in changes in areas E1 and K1 such a conclusion would be irresponsible.

In order to analyse the perception of security, studies on public opinion also resort to questions about leaving home after dark (see Table 6.6), making it possible to accurately measure the scale of threats perceived by the respondents as affecting their actual behaviour (in this case, leaving their house or flat). This particular study also benefited from posing such questions. As in the case of other indicators of feeling safe, this one also improved among the population of all four areas. Differences between the experimental and control areas are ambiguous. In the first pair of areas the number of residents who felt safe and very safe after nightfall increased significantly (from 40.2% to 51.5%). A larger increase was, however, manifested in feeling secure in the control area (K1). In the second pair, as in the two previous questions about feeling safe, there occurs a significant difference between the two areas. Residents in the control area felt more secure in 2007 than in the previous year, but only by 4.1%, while the perception of safety in the experimental area grew by 37.3%. As in earlier cases, these results do not demonstrate the preventive efficacy of public video surveillance systems.

RESULTS OF THE WARSAW SURVEY VERSUS WIDER TRENDS

Undoubtedly, each area featured a reported change (decline) in the number of crimes whose victims were residents and their children. In addition, the subjective evaluation of feeling safe improved significantly in all areas. As has been noted, differences recorded between control and experimental areas cannot be proof of the effectiveness of public video surveillance systems. The only conclusion is that the expected results have not been achieved. These results were defined as a decrease in crime (and an increase in feeling safe) in the experimental areas that is larger than in the control ones. Not knowing in which areas the cameras had been installed, and basing only on obtained relevant research results it would be impossible to determine the area in which the variable (camera) had been introduced.

Despite the mentioned limitations of police statistics and the relatively small sample size, one can attempt a comparison in order to obtain results in the course of research. In 2007 the number of crimes recorded by the Warsaw Metropolitan Police fell in comparison to the previous year by more than 19,000 (from 108,168 to 88,916; Warsaw Metropolitan Police 2011). This figure represents a 17.8% decline in the total number of offences registered by the police. The scale of the change is comparable to the one in the E1 area (a drop of 23.1%), but remains considerably smaller than the one recorded in three other areas (K1 at 47.4%; E2 at 65%; and K2 at 69.1%). At the same time, the decrease in the victimisation of respondents in Warsaw was much higher (rising from 37.7% to 71.4%) than was indicated by police statistics. This may be proof of a significant number of offences undisclosed in police statistics, which would be revealed by victimisation research. According to a Polish study on the dark figure of crime, only 47% of crimes are (for various reasons) reported and recorded by the police (Polish Ministry of the Interior 2011). This fact may have a significant impact, especially in the case of crimes perceived by the victim as trivial; the absence of statistics translates into differences in the recorded changes (if the latter concern offences not reported to the police). All complaints and alleged differences do not alter the fact that this trend, which highlighted the results of research conducted in the relatively small area of Warsaw, is part of a broader tendency apparent both upon the basis of figures from the Warsaw Metropolitan Police headquarters and the entire country, where the number of criminal offences listed in police statistics has fallen from 893,389 in 2006 (Polish National Police 2007) to 749,317 in 2007 (Polish National Police 2011), a decrease of 16.1%. The most likely reasons for the fact that we are dealing with large-scale changes are demographic transformations (an aging population) and mass-scale emigration.

It is interesting to compare the obtained results with those of two Polish editions of Research on Crime. A consortium of three research centres (Centrum Badania Opinii Społecznej—CBOS, Pracownia Badań Społecznych—PBS DGA, and Ośrodek Badania Opinii Publicznej—TNS OBOP) carried out such a Research on Crime twice: in January 2007 and January 2008. The few months' difference between the two phases of the research results may provide material for comparison. The most noteworthy is the outcome achieved in the Warsaw Metropolitan Police area of operation, where respondents over fifteen years of age were questioned (a significant difference in relation to research featured in this study, which includes only adults). The sample consisting of 1,000 respondents was representative for the Warsaw Metropolitan Police area (not just the region of

Warsaw, but also the adjacent boroughs), which distinguishes it from the sample representative for the studied areas. Despite these limitations, it is worthwhile to become familiar with the survey outcome.

In the course of research conducted in January 2007, 62.1% respondents declared that they felt safe when walking after dark near their place of residence. One year later, this group grew to 74.3% (Polish National Police 2008). The change (calculated according to the scheme adopted in the research) is 19.6%—that is, more than two times smaller than the one registered in the three studied areas (a slighter change occurred only in K2). The four areas under research differed from the Warsaw Metropolitan Police area, showing a relatively lower perception of safety during the first phase of the study, which influences the recorded amendment; it turns out, however, that the trend observed in the surveyed areas is not an exception; the feeling of security grew not only at the national level (from 70.1 % to 75.5%), but also in the Warsaw Metropolitan Police area. If the increased perception of security has been recorded to such an extent, it is difficult to assign this effect to public video surveillance alone, just as in the case of demographic change and emigration. After Poland joined the EU in May 2004, 1.5–2 million people—mainly young citizens, who are statistically more likely to become offenders and crime victims—emigrated from Poland to find better-paying jobs. Demographic changes on such a scale (the population of Poland totals about 38 million, so 2 million is equal to 5%) always influence crime data; the change in question, however, had actually started two years prior to the first phase of the research.

All the findings from the Warsaw survey confirm the results obtained in the briefly summarised UK, German, Austrian, and Spanish evaluations. There is (so far) no scientific proof that public video surveillance installation acts as a crime deterrent or that it makes people feel more secure (even if they are certain that it will). Even the limitations of the Warsaw survey (the size of the research sample range and only four research areas) do not create an obstacle for concluding that the heart of the matter does not lie in the climate or culture, erroneous methodology, or biased researchers; even sophisticated technical systems of cameras monitored by trained and well-paid operators do not prevent crime.

It is important to stress that regardless of the methodology chosen by researchers from different countries it has not affected the main finding: public video surveillance does not deter offenders from committing crimes in the area covered by cameras. Nor has it made people (residents, passers-by) feel more secure. To some extent, even if similar results are obtained in several cities in different countries facing various crime problems and with dissimilar levels of the perception of security, we may conclude that the observed trend is not a mere mistake or a social artefact. There is no causality between functioning public video surveillance and altering crime rates, or between the presence of cameras and the perception of security. The gap in scientific data on the effectiveness of public video surveillance systems, present at the beginning of the 1990s, has been filled with the results of several independent surveys.

Paradoxically, it was expected that the scientific world would legitimise the link between the 'silver bullet' and crime. Although at the onset of the rise of video surveillance the biggest source of funding the systems was public money, and many existing systems are still financed by the public budget, there were no data legitimising it. Popular belief reinforced by 'marketing data' was sufficient for decision makers all over Europe to finance this particular crime prevention tool. Today, when there are several available results and even public opinion seems to have changed its attitude towards video surveillance, it seems that people responsible for security choose one of three reactions, the first being a denial that such results exist or a tendency to undermine their credibility. The second tactic, which could be labelled as a 'never-ending story', concentrates on improving the system—purchasing more sophisticated equipment (cameras, software) and training system operators. This approach is very similar to that of the 'arms race': there will always be new (better) equipment available if there is on-going demand for it. The third tactic appears to be more adaptive. Results denying the crime deterrent value of public video surveillance systems are acknowledged, but new ways of system utilisation are being discovered. More and more video surveillance systems are being presented as providing valuable information for law enforcement agencies and evidence for court proceedings. It seems that the original and main justification is becoming increasingly less evident and will be replaced by a new explanation, albeit within the same area—that is, combating crime—even in a situation wherein Scotland Yard declared that the public video surveillance system helps to solve only 3% of street robberies and called it "an utter fiasco" (Bates 2008). If sufficient elements of the public video surveillance system are already functioning, it is very unlikely that it will be shut down, regardless of the outcome of its evaluation and declared effectiveness (unfortunately, in terms of the effective use of public funds). The examples of the assessed systems in the UK, Hamburg, Vienna, Malaga, and Warsaw, together with a lack of other reactions to research results questioning the systems' achieved objectives, prove the veracity of this assumption.

REFERENCES

Bibliography

Bates, Daniel. 2008. "Billions Spent on CCTV Have Failed to Cut Crime and Led to an 'Utter Fiasco', says Scotland Yard Surveillance Chief." *Daily Mail*, May 6.

Cerezo Dominguez, Ana I., and José Diez Ripolles. 2010. "La videovigilancia en las zonas publicas: su eficacia en la reduccion de la delincuencia." *Boletin Criminologico* 121: 1–4.

Feliksiak, Michał. 2011. *Poczucie bezpieczeństwa, zagrożenie przestępczością i stosunek do kary śmierci. Komunikat z badań CBOS*. Warsaw: CBOS.

Gill, Martin, and Angela Spriggs. 2005. *Assessing the Impact of CCTV*. Home Office Research Study 292. London: Home Office.
Heinz, Wolfgang. 1998. "Kriminalprävention—Anmerkungen zu einer überfälligen Kurskorektur der Kriminalpolitik." In *Entwicklung der Kriminalprävention in Deutschland*, ed. H.-J. Kerner, J.-M. Jehle, and E. Marks. Bonn: Forum-Verlag Godesberg, 17–60.
Newburn, Ted. 2007. *Criminology*. Cullompton, England: Willan.
Miller, Joel. 2007. "Evaluacion de la videovigilancia en Malaga: El diseno de un quasi-experimento." *Boletin Criminologico* 94: 1–4.
Norris, Clive. 2010. "Closed-circuit Television: A Review of Its Development and Its Implications for Privacy." In *International Handbook of Criminology*, ed. Shlomo Giora Shoham, Paul Knepper, and Martin Kett. Boca Raton, Fla.: CRC Press, 395–423.
Pearson, Geoffrey. 1983. *Hooligan: A History of Respectable Fears*. London: Palgrave Macmillan.
Rothman, Robert. 2010. "Sicherheitsgefühl durch Videoüberwachung? Argumentative Paradoxien und empirische Widersprüche in der Verbreitung einer sicherheitspolitischen Maßnahme." *Neue Kriminalpolitik* 2010(3): 103–7.
Sherman, Lawrence W. 1998a. "Conclusion: The Effectiveness of Local Crime Prevention Funding." In *Preventing Crime: What Works, What Doesn't and What Is Promising: A Report to the United States Congress*, ed. L. W. Sherman, D. Gotfredson, D. MacKenzie, J. Eck, P. Reuter, and S. Bushway. College Park: University of Maryland, 680–715.
Sherman, Lawrence W. 1998b. "Preventing Crime: An Overview." In *Preventing Crime: What Works, What Doesn't and What Is Promising: A Report to the United States Congress*, ed. L. W. Sherman, D. Gotfredson, D. MacKenzie, J. Eck, P. Reuter, and S. Bushway. College Park: University of Maryland, 3–7.
Siemaszko, Andrzej. 2001. *Kogo biją, komu kradną. Przestępczość nie rejestrowana w Polsce i na świecie*. Warsaw: Oficyna Naukowa.
Welsh, Brandon C., and David P. Farrington. 2002. *Crime Prevention Effects of Closed Circuit Television: A Systematic Review*. Home Office Research Study 252. London: Home Office.
Zurawski, Nils. 2007 *Videoüberwachung in Hamburg. Abschlussbericht*. Hamburg: Institut für kriminologische Sozialforschung der Universitat Hamburg.
Zurawski, Nils. 2010. "'It Is All about Perceptions': Closed-circuit Television, Feelings of Safety and Perceptions of Space—What the People Say." *Security Journal* 23: 259–75.

Websites

Crises Management and Safety Department, City of Warsaw. 2012. Public Video Surveillance Bureau website. Accessed 10 February 2012 at http://www.zosm.pl/index.html.
Polish National Police. 2007. *Raport statystyczny 2006*. Accessed 21 December 2011 at http://www.policja.pl/portal/pol/1/5485/.
Polish National Police. 2008. *Strachu coraz mniej*. Accessed 21 December 2011 at http://www.policja.pl/portal/pol/1/18437/Strachu_coraz_mniej.html.
Polish National Police. 2011. *15 przestępstw mniej co godzinę*. Accessed 21 December 2011 at http://www.policja.pl/portal/pol/1/13255/15_przestepstw_mniej_co_godzine.html.
Polish Ministry of the Interior. 2011. *Badanie Poczucia Bezpieczeństwa, Oceny Pracy Policji oraz Ciemnej Liczby Przestępstw 2007–2009*. Accessed 21 December 2011 at http://razembezpieczniej.mswia.gov.pl/download.php?s=23&id=204.
Warsaw Metropolitan Police. 2011. *Bezpieczeństwo w Warszawie*. Accessed 21 December 2011 at http://www.ksp.waw.pl/?.

7 From Privacy Protection towards Affirmative Regulation
The Politics of Police Surveillance in Germany

Eric Töpfer

INTRODUCTION: PRIVACY AS PUBLIC GOOD

While leading scholars of surveillance studies denounce privacy as 'hyper-individualistic concept' which is not only blind to the wider societal and political implications of surveillance but also helpless, and therefore useless, against its excesses (Monahan 2006: 23; Stalder 2002; Lyon 2003), other authors argue that "privacy serves not just individual interests but also common, public, and collective purposes" (Regan 1995: 221). Bennett and Raab (2006), drawing on Alan Westin's "four states of privacy",[1] counter the view that privacy and social values are necessarily antithetical, noting that "intimacy and anonymity imply the ability of individuals to engage others, rather than signifying their withdrawal from society. These two 'states' therefore sustain participation in collective political life, including such modes of activity as associating politically with others or voting without fear of surveillance. As seen in analyses such as these, this relationship between privacy and political participation opens an avenue, even within the conventional [i.e. liberal] paradigm, for considering privacy as a value for society beyond the single individual or beyond a simple aggregate of individuals" (Bennett and Raab 2006: 24).

From the perspective of surveillance studies these arguments are an implicit reminder that total surveillance not only aims to expose those under scrutiny but also to silence them. Bentham's and Foucault's imagined inmates of the ideal panoptic prison are subjected to the disciplining gaze of an unseen watcher and completely isolated in their cells, being cut off from all (visual) contact with their neighbors: "The crowd, a compact mass, a locus of multiple exchanges, individualities merging together, a collective effect, is abolished and replaced by a collection of separated individualities" (Foucault 1977: 201). These visions of total surveillance culminate in 'silent' prison systems such as nineteenth-century Auburn or twenty-first-century Camp X-Ray where the inmates are not only incarcerated in 'transparent' cages but are also strictly prohibited to communicate with their fellow prisoners (Winkelmann and Förster 2007). Therefore surveillance, at least in its panoptic version, is not only a means to produce docile individuals but also to inhibit the building of what Bourdieu (1983) and others call "social capital", the actual and potential power resources derived from social relations, group affiliation, and enduring networks of kinship, friendship, association, or class. As such, surveillance may not only target individuals' 'right to be let alone' but also their ability to interact with others and to participate in social and political life, while privacy may protect individuals from isolation and society from atomization and thus serve the public interest.

"[T]urning the discussion around to emphasize the social importance of privacy", Regan suggests, "will have important policy implications". She expects that such a turn would provide privacy advocates with a stronger basis upon which they can build their arguments against the expansion or abuse of surveillance regimes: "The policy debates will be different if the policy issue involves the balancing of two societal interests. In turn, the policy outcomes may also be different" (Regan 1995: 231).

In this chapter I will question this expectation by examining the politics of police surveillance in Germany, where constitutional law has established a concept of (information) privacy that also explicitly values its importance for political participation and democratic society. I will, first, briefly introduce the scope and limits of this concept, the *Recht auf informationelle Selbstbestimmung* (right to informational self-determination). Second, I will show how the establishment of this new civil right has framed legislative policies in the area of *innere Sicherheit* (internal security) by tracing the central developments in this policy area from the late 1970s till today. Though necessarily painting with a broad brush, this aims to demonstrate that the effect of the right to informational self-determination was detailed but affirmative legal regulation of increasing surveillance powers rather than their effective containment, thus preparing the ground for a neoliberal vision of freedom as mere individualism. Third, I will exemplify this thesis by a more detailed account of how police closed-circuit television (CCTV) evolved in the German arena of electoral politics and policing. Against Regan's optimistic assumption I will then show how powerful advocacy coalitions were able to colonize the 'public interest' in order to justify the expansion of CCTV surveillance despite the particular German notion of information privacy. I will conclude, however, that the Federal Republic's legislative sleepwalk into a new police state is obsolete as excessive surveillance does not match the political economy of German policing. In the case of CCTV, reluctant decision makers within the police force and scarce financial resources have limited its expansion, at least in comparison to 'maximum surveillance societies' (Norris and Armstrong 1999) such as Britain's.

THE 1983 CENSUS DECISION: THE MAGNA CHARTA OF THE INFORMATION AGE?

When in the early 1980s—ten years after the count in 1970—the details of the forthcoming fourth census of the German Federal Republic were

From Privacy Protection towards Affirmative Regulation 173

unveiled, the plans were quickly facing massive resistance not only from the new social movements[2] but also from such prominent people as the author Günter Grass; Helmut Simon, judge on the Federal Constitutional Court; and Manfred Güllner, founder of the influential public opinion poll research institute Forsa. The critics argued that the detailed questionnaires, in connection with new central databases and the processing power of mainframe computing, would allow to trace back the identity of the interviewed. They feared the *gläserner Bürger* (transparent citizen), or even a decisive step towards an Orwellian surveillance state, and instituted proceedings against the Census Act at the Bundesverfassungsgericht (Federal Constitutional Court). Given the suit, the census—planned for April 1983—was first postponed and then suspended, after the court overruled central sections of the Census Act on 15 December 1983.[3] In its seminal decision the court derived the right to informational self-determination from the right to personality as protected by the second article (personal freedom) in connection with the first article (human dignity) of the (Western) German quasi-constitution, the Grundgesetz (Basic Law). "The fundamental authority of the individual to decide by her or himself if and in which context to unveil issues of personal life", the court ruled, "requires significant protection under the actual and future conditions of automated data processing".[4] Given the technical potential for an unlimited storage of personal data which were beyond the control of the data subject and being accessible from a distance, and given the potential to assemble personal profiles from these data, the court saw—in contrast to recent manual filing systems—not only personal freedom but also democracy under threat:

> Whoever expects that e.g. the attendance of an assembly or the participation in a civic action group will be registered by the authorities and that this will probably cause risks, may probably abandon their corresponding fundamental rights (Art. 8,9 Basic Law) [freedom of assembly and association]. This would not only impact the individuals' chances for development but also the public interest because self-determination is a necessary condition for the functionality of a liberal democratic polity which is based on its citizens' ability to act and to participate. (Translation from 1983 population census verdict by the Federal Constitutional Court, as quoted in: Weichert 1998: 70)

The Constitutional Court declared, in principle, that each person possesses the right to control her or his representation via data shadows (the sum of all traces of information that an individual leaves behind through digital activities) and that therefore the collection, processing, usage and transfer of personal data require informed consent by the data subjects. The decision is seen as a milestone of data protection. However, the court did not establish an unlimited warrant of the newly declared right, noting, "The individual has not a right over his or her data in the sense of an absolute and unrestricted authority; rather he or she is a socially embedded personality depending on communication. Information, also personal information, represents a simulacrum of social reality which does not only refer to the affected person. . . . Therefore the individual has to accept restrictions of his or her right to informational self-determination in the prevailing general interest".

According to the court, restrictions of the German version of information privacy require a legal basis and have to comply with the principle of proportionality. Moreover, they have to be clear and transparent for those who are affected, meaning that the overall process of data processing should be understandable and traceable, even if this does not imply any form of intervention by the data subjects. Any collection of data 'by force'—that is, without informed consent—requires that the purpose is "specifically and precisely determined by the legislator". The data have to be appropriate and necessary to meet the declared purpose; any storage of personal data for unclear purposes would be illegitimate, as would any transfer to third parties without a legally determined purpose.

However, the decision applied to public data collection for statistical purposes (Bäumler 2001) stresses that police and other security agencies do also fall under the scope of the census decision. As their use of information may entail particular "intimidation" affecting personal freedom and democratic participation (Bäumler 2001: 746), they should respect information privacy with specific care. Nonetheless, the litmus test for the value of an institutionalized concept of information privacy that is also sensible for its societal importance is not constitutional law but the law in practice. To approach the question of how useful it is in containing surveillance, I will now examine the development of police surveillance within the German framework.

LEGALIZING THE NORMALIZATION OF THE STATE OF EXCEPTION: POLICIES OF INNERE SICHERHEIT IN THE WAKE OF THE CENSUS DECISION

Taking a course towards the "preventive turn" (Narr 1998), or what is discussed in the Anglophone literature as the 'master shift' towards proactive policing of risk societies and actuarial justice (Cohen 1985: 127; Ericson and Haggerty 1997; Feeley and Simon 1994), can be dated in the case of the Federal Republic of Germany to the late 1960s. In his 1969 government statement Chancellor Willy Brandt, heading the first coalition of the Social Democratic Party (Sozialdemokratische Partei Deutschlands, SPD) and the liberal Free Democratic Party in post-war Germany, announced the modernization of policing, implicitly responding to significant social change and political turmoil of that years. His announcement materialized when the Conference of the Ministers for the Interior (Innenministerkonferenz) presented their first Program for Internal Security (Programm für die Innere Sicherheit) in 1972, amidst the escalating militant struggle of the first generation of the left-extremist urban guerilla Red Army Faction (Rote Armee Fraktion, RAF) against the German Federal Republic and shortly before the bloody Palestinian assault of the against the Israeli athletes at the Olympic Summer Games in Munich that became known as Black September (Narr 1994; Kunz 2005). The central aim of this program was to prepare the "democratic state" against "the rise of crime and the brutalization of political forms of expression" by reforming the organization and agenda of the internal security agencies of the Federal Republic (Busch et al. 1985: 230): it was decided that police personnel should be increased, their training harmonized, research and development in the fields of forensic science and police equipment professionalized and centralized, and new tasks and police powers established.

The achievement of most of these detailed objectives fell in West Germany's federal polity under the jurisdiction of the *Länder* (states) and was realized by subsequent amendments of several state police acts. The new stop-and-search powers of the 1978 Musterentwurf für ein einheitliches Polizeigesetz (Model Draft for a Uniform Police Act) that guided most of the police act amendments throughout the late 1970s and '80s marked a first significant breach with traditional police law. While the powers of the police to stop and search citizens in public places without the need for individual suspicion (or probable cause) was previously limited to events of danger *in concreto* or *in actu*, the relevant sections of the Model Draft indicate the shift towards proactive policing, or in German terms, from *Gefahrenabwehr* (defense of danger) towards *Gefahrenvorsorge* (provision for danger), being inserted between traditional public order policing and criminal investigation. Through the legal figure of *gefährliche Orte* (dangerous places) and *gefährdete Objekte* (premises at risk), at or nearby which the police became authorized to stop people for identity checks and search those who are unable to identify themselves, the submission to police control was transformed from an exceptional infringement of civil rights into a duty of the general public, reversing the presumption of innocence. "The previous legal state of emergency was turned into legal normality", conclude Busch et al. (1985: 200). However, this normalized state of exception declared by arbitrary designations of locations as 'dangerous' by the police was limited in space and time, and was therefore of minor scope and reason for public concern. "The governmentality of unease", as Bigo notes, "increases the exception and banalizes it" (2006: 47).

Many other exceptional tactics of proactive policing, though, evaded formal legalization and were instead practiced throughout the 1970s in an unregulated 'grey area'—as, for example, the first *Rasterfahndung* (dragnet investigation) aimed at locating RAF members. Consequently, the realm of policing, law enforcement, and national security was excluded from the scope of the first Federal Data Protection Act (Bundesdatenschutzgesetz, BDSG), passed in 1977, as (in)security experts feared a paralysis of police and intelligence services work when fair information principles would apply to the collection and processing of personal data in this field.

This situation significantly changed with the advent of the right to informational self-determination through the Census Decision. As elaborated above, the Constitutional Court indisputably ruled that any collection and use of any personal data constitutes an infringement of civil rights and is, therefore, only legitimate provided informed consent by the data subject, or in the "prevailing general interest" and under a regime of legal regulation.

In the years following 1983 a wide range of legislative activities were tasking German parliaments as a response to the Census Decision. These aimed, similar to legislation guided by the first Model Draft for a Unified Police Act, at amending existing acts in order to explicitly legalize surveillance practices and technologies already deployed by the police, the secret services, and other public authorities involved in policing—most notable of these at the federal level was the amendment of the Versammlungsgesetz (Assembly Act) in 1989 which regulated, among other things, the deployment of police cameras against public protest, and the passage of the euphemistically named Gesetz über die Fortentwicklung der Datenverarbeitung und des Datenschutzes (Act for the Further Development of Data Processing and Data Protection) in December 1990, which revised and created, in one swoop, both the Federal Data Protection Act and acts on the federal secret services, respectively.[5] At the state level, following a revision of the Model Draft for a Unified Police Act, long-anticipated practices such as CCTV surveillance in the vicinity of 'premises at risk' or targeting apolitical crowds—for example, football supporters,—were legalized.

Moreover, the roll-out of new measures, desired by security professionals, was legally prepared such as the introduction of the machine-readable national ID card in the late 1980s by the amendment of the Personalausweisgesetz (Act on Personal ID Cards) in April 1986. This process, for which Narr (1998) coined the term *Vorwärtsverrechtlichung* (forward regulation/ legalization), was accelerated after the end of the Cold War. While 'terrorism'—though transforming from the personalized Baader-Meinhof Gang into a faceless and global 'guerilla diffusa'—was the uncontested leitmotif in discourses on internal (in)security and of respective legislation throughout the 1970s and early '80s, this motif was supplemented and overshadowed by drug trafficking, 'organized crime', criminal foreigners and street crime from the late 1980s on and in particular after the fall of the Iron Curtain and German unification. All these motifs merged more recently with the events of 11 September 2001, after which dealers selling heroin or cocaine in the streets of Berlin, Hamburg, or Frankfurt, though seen as small-time criminals, were held responsible for fuelling the war economies in 'failed states' which breed and shelter globalized terrorism.

It is impossible to comprehensively list the series of new security acts that were breathlessly passed in German parliaments after 1990, but just a few milestones at the federal level include the OK-Gesetz (Organized Crime Act, 1992), the Bundesgrenzschutzgesetz (Federal Border Police Act, 1994), the Gesetz zur Verbesserung der Bekämpfung der OK (Act to Improve the

Combat against Organized Crime, 1998),[6] the Bundeskriminalamtgesetz (Federal Criminal Office Act, 1998), Schily-Pakete[7] I and II (Anti Terror Acts) in the aftermath of 9/11, and, most recently, the Vorratsdatenspeicherungsgesetz (Data Retention Act, 2007) implementing the European Directive 2006/24/EC and authorizing six-month storage of connection data of telecommunications (see, among others, Busch 1994; Roggan 2000; Normann 2007; Leutheusser-Schnarrenberger 2008).

Central parts of all these acts are affirmative regulations of old practices or new powers for the police and intelligence services to collect, process, and share data. They "encapsulate" (Narr 1994: 11) information privacy into police and internal security law. Rather than limiting surveillance or submitting it to fair information principles, they do, as far as Bäumler (2001) is concerned, reverse the objective of the Census Act to establish clear standards for the use of personal data "by perfectionist meticulousness into its opposite" (749). The host of legal instruments, developed since the 1970s, normalized 'minor' states of exception and expanded their scope in time and space. Kutscha (2001) argues that this development directs towards an "overregulated absence of regulation" qualified by vague legal terms and a complex and confusing set of cross-references which, instead of submitting police action to due process, aims to keep as much options open for it as possible. Therefore, Roggan (2000) concludes, Germany is, in legal terms, carelessly moving from the rule of law towards a police state. Given the forward legalization of expanding surveillance powers for security agencies, Busch (2006) notes that the interventions by German data protection authorities are increasingly limited to the persecution of abuse while fundamental critique of their use is escaping their competence. The Federal Constitutional Court, though, has recently drawn clear limits to restrictions of information privacy by police surveillance with decisions on electronic eavesdropping in private homes, on surveillance of telecommunications, and on online computer searches. However, Busch argues, the essence of these decisions is "bedroom data protection", as they only protect the so-called *Kernbereich privater Lebensführung* (core of private life) and therefore implicitly redefine public space from a forum of democratic interaction into a "dangerous place" which has to be put under surveillance in the "public interest".[8] From this perspective, we indeed witness the rise of a "hyperindividualistic" neoliberal version of privacy which represents the "tyranny of intimacy" that does not only distort society but also frustrates and corrupts the individuals themselves (Sennett 2000). But the problem results not from the concept of privacy per se but from the colonization of the 'public interest' by powerful coalitions advocating the expansion of surveillance in the name of a newly declared 'civil right to security'.

To illustrate the process of how such coalitions colonized the 'public interest' in a more detailed retrospective, I will examine next the rise of open-street CCTV in Germany. Having in mind the incremental and contested character of the implementation and adaptation of new technologies by the police (Heinrich 2007), I will not only look at the level of legislation but also at how and why the police acted within the emerging institutional framework.

THE LIMITED RISE OF OPEN-STREET CCTV

The deployment of police cameras in public space has been practiced in the German Federal Republic for fifty years to manage traffic flows, control crowds, or combat serious crime such as bomb attacks by the Red Army Faction (Weichert 1988). In general these efforts targeted specific gatherings such as sporting events or political demonstrations and were thus of limited duration; when the efforts were more permanent they were not meant to collect images of individual persons, as in the case of traffic management. The permanent CCTV surveillance of public areas for purposes of crime control became an issue only in the mid-1990s.

In April 1996, the city of Leipzig started a four-week trial with police video surveillance to monitor the area in the vicinity of the central railway station. At the heart of the measure were efforts to disturb and displace drug-related street crime in the city center. Around the same time, the police on the island of Sylt utilized a surveillance camera to monitor a pedestrian zone in the seaside resort Westerland during the high season in order to protect tourists from disturbances by loitering youth. While the Leipzig system was dismantled in the late 1990s, the city became a mecca and reference point for open-street CCTV in Germany. After the pilot project was declared a success by the local police department, the temporary installation was made permanent and expanded (Müller 1997, 2000; Töpfer 2005).

Within a few years other cities in eastern Germany followed the example of Leipzig: Up to the end of 1999 police departments in Magdeburg, Dresden, Halle, and Dessau set up small schemes with not more than three cameras. Although a local administrative court waived a legal challenge against the system in Halle, it soon became apparent in these early days of open-street CCTV that the police departments were, given the census decision, skating on thin legal ice. Police officers and security administrators discussed in professional journals the quality of CCTV images as personal data. Some argued that images from camera monitor systems without recording or from cameras that did not zoom in lacked such quality. Others claimed that the deployment of CCTV should be allowed under the general clause of German police acts that authorized the police to make use of appropriate means in order to fulfil their task. or pointed to the Code of Criminal Procedure, which, revised by the Organized Crime Act in 1992, authorized the police to deploy even covert video surveillance in order to investigate crime (for these arguments, among others, see Keller 2000; Roos 2002; Roll 2003; Brenneisen 2003). However, eventually the demand for legal certainty of administrative action prevailed, and the bureaucracies of several ministries of the interior began to draft bills aiming to achieve affirmative regulation of open-street CCTV,[9] and Saxony amended its police act in 1999, tailoring it to the requirements for CCTV surveillance practices in Leipzig and Dresden.

These early pilot projects of open-street CCTV were embedded in the changes to urban landscapes and new discourses on crime and (in)security in cities. The decline of industry and the rising importance of service and consumption shifted the focus of urban economies and the interests of local elites. Inner-city areas were reinvented and refurbished to re-create a public image and attract tourists and consumers to newly developed shopping and entertainment districts as well as train stations. The rediscovery of city centers brought diverse conflicts over space and the development of new policing strategies: public drinking, drug dealing, and 'aggressive' begging were depicted as menaces to public order. Conservative politicians advocated 'reclaiming the streets' for the 'decent people' and efforts were undertaken to criminalize and exclude unwanted behavior from downtown areas by local by-laws. Roundtables on crime prevention established new networks of social control at the local level, and new instruments for the policing of places designated as 'dangerous' penetrated police law and practices. In this broader context open-street CCTV began to attract the interest of the police, local businesses, and politicians as both a promising "instrument for the efficient deployment of the scarce resource police"[10] and as an allegedly effective tool for crime prevention.

Thus, open-street CCTV entered the arena of electoral politics. During the campaign for the national election in 1998, the dawn of the chancellorship of Helmut Kohl, several state branches of the Christian Democratic Union (Christlich-Demokratische Union, CDU), Germany's conservative party, made crime in the city an issue and propagated open-street CCTV as a panacea for crime prevention. Though—or because—the election ended in disaster for the Christian Democrats, CCTV remained on the political agenda. On top of the crisis of the CDU that followed the defeat of Helmut Kohl, the conservative candidate for the elections in Germany's largest state North Rhine Westphalia, Jürgen Rüttgers, presented in spring 2000 a paper on public safety that emphasized the alleged benefits of public-area CCTV. A few weeks later, in May 2000, the Conference of the German Ministers for the Interior, following an initiative by the CDU-led southern state of Baden-Württemberg, unanimously declared "the overt deployment of video surveillance at crime hot spots in public areas an appropriate instrument to support the realization of police tasks in the context of crime prevention and criminal investigation" (Innenministerkonferenz 2000). The preparation of this declaration had begun in autumn 1999 when the Working Group II Innere Sicherheit of the conference was assigned to draft an opinion on CCTV. In charge of drafting this opinion was a high-ranking police officer from the state of Schleswig-Holstein who was responsible for the police cameras in Westerland/Sylt in 1996. With their declaration, the ministers for the interior ignored the warning of the 59 Conference of the German Data Protection Officers, who, being alerted to the developments, met two months earlier in order to declare that all people have the fundamental right to move in public areas without their behavior being recorded by cameras.

The decision of the ministers kicked off a new series of police act amendments aiming to legalize open-street CCTV at crime hot spots. Only a few days later, the state assembly of North Rhine Westphalia passed a new regulation. Challenged by Jürgen Rüttgers, a coalition government of the SPD and the Green Party authorized CCTV at crime hot spots. In the same year similar amendments were also passed by governments led by either the CDU or the SPD in Saxony-Anhalt, Brandenburg, Baden-Württemberg, and Hessen. Bavaria, Saarland, Bremen, and Lower Saxony followed in 2001; Thuringia in 2002; Rhineland Palatine in 2004; Hamburg in 2005; and Schleswig-Holstein in February 2007. In Mecklenburg-Western Pomerania, a coalition of Social Democrats and Socialist Part members refused to pass any such legislation until it amended the police act in the advent of the G8 summit in Heiligendamm. In 2008 only Berlin, governed by a coalition of Social Democrats and Socialist Part members explicitly refused to legalize open-street CCTV.

However, not all state police made immediate use of their new power. Permanent systems are only in operation in nine of the sixteen German states. The police in Rhineland Palatine, for example, have so far only deployed cameras during special events—for example, a 200-camera network during the FIFA World Cup 2006 in the venue city of Kaiserlautern. By 2008, police departments in about thirty cities operated open-street CCTV systems on a permanent basis. In total these systems have about 100 cameras; the largest network is currently installed in Hamburg, with seventeen cameras.[11] Though all these systems can record video footage, not all of them do so on a permanent basis as the provisions on recording practices and storage times for footage differ from police act to police act. Most systems are owned and operated by the state police, but some operate in collaboration with municipal *Ordnungsämter* (public order departments) and/or the Bundespolizei (Federal Police),[12] which is in charge of policing the national railway network and train stations, among other duties.

In several cases the operation of surveillance systems was stopped or paused due to various reasons. In Bielefeld, the pilot project for North Rhine Westphalia, a small but influential group of privacy advocates initiated critical debate and mobilized public awareness, and the project was halted after being declared a success that made further operation obsolete. However, it again started operation after a new amendment of the state police act weakened the threshold for the qualification of areas as crime hot spots in 2004. In Stuttgart, Heilbronn (Baden-Württemberg), and a few other cities the operation of systems was ceased after the police concluded that their tools were successful and thus saw no legal basis for further operation as the locations under surveillance were not crime hot spots any more. In Weimar, a system was dismantled after three days in operation because it was discovered that the cameras were also monitoring the entrance to the editorial office of a local newspaper. However, in the majority of cities the police continued to run their systems even after they were declared a success. In several cities, including Hamburg and Munich, the networks were expanded after a while and many municipal councils declared their intention to put their city centers under surveillance, provided that the police agreed. A slowdown of the expansion of open-street CCTV is, therefore, not in sight.

The moderate but steady expansion in surveillance is driven by a coalition of internal security forces. Though such surveillance has largely been initiated by the Christian Democrats, the Social Democrats have also strived to present their party as the protector of law and order. They have supported the amendment of police acts as lead partner in big coalitions with the Christian Democrats in Brandenburg and Bremen. In North Rhine Westphalia they put their junior partner, the Green Party, under intense pressure to vote for open-street CCTV, and in Saxony they even voted, together with the Christian Democrats, against their junior coalition partner, the Party of Democratic Socialism for the revised police act. Although their youth organization opposes expanding surveillance, the liberal Free Democrats supported as a junior partner in coalition governments the legalization of open-street CCTV. The Left Party (former the Party of Democratic Socialism) has been the only party to regularly oppose CCTV surveillance at crime hot spots, but could not resist, as a member of the ruling coalition government, voting for the police act aimed at preparing the ground for G8 summit security in 2007 in the northern state of Mecklenburg-Western Pomerania.

At the local level, party affiliations do not determine attitudes towards open-street CCTV. Mayors and local councilors from most major parties have voted for the use of CCTV and are lobbying their local police departments to agree. Their concerns are often for city marketing, and they are supported by local chambers of commerce or other business associations that aim to attract tourists and affluent consumers by improving the image of shopping and entertainment areas. Moreover, the electronic industry was busy in lobbying the decision makers at both the police and the political levels in the kickoff phase around the year 2000. Anticipating privacy sensibilities, the marketing strategy of their major association ZVEI[13] was offering, on top of the basic surveillance equipment, "data protection through technology", also known under its euphemistic name "privacy enhancing technology". Moreover, they were developing concepts for a pilot scheme at a prominent location in Berlin, which only failed as the local Christian Democrats, who were supporting the plan, were voted out of power after a major scandal on public finances.

The legalization and implementation of open-street CCTV is usually justified with reference to opinion polls which allegedly find a significant majority of fearful citizens in support of its use. However, social research demonstrates that most people are rather uninformed, and it makes clear that their opinion is shaped by hegemonic discourses on CCTV and its crime effect (Klocke and Studiengruppe 2001). Moreover, research suggests that public acceptance is not supported by fear of crime. Instead, studies have found that the belief in the myth of CCTV as a panacea against crime, trust in the state and its police, and desire for social order to be much more decisive factors determining the positive attitudes of CCTV supporters (Reuband 2001; Hölscher 2003). Therefore, public support for the expansion of CCTV in Germany can also be explained in the broader context of what Heitmeyer (2008) calls "group-related misanthropy": with the rising experience of social disintegration and precarious lives, Heitmeyer and his team proved in an on-going long-term study, envy and scapegoating are increasingly penetrating the social fabric. Thus, jealous defense of established status and indifference at the fate of stigmatized groups turns out to be the other side of the coin of neoliberal responsibility. In such a climate of fear, the promise of security and restoration of order by excluding those who are already at the margins of society is a tempting option which professional politicians who want to succeed in electoral politics can hardly resist.

Given the powerful advocacy coalition of policemen in search of new tactics, security industries in search of new markets, downtown businesses in search of a competitive advantage, and politicians in search of votes, and their seduction of a an electorate in search of order, it is not surprising that critics of CCTV are marginalized. Data protection officers and civil libertarians who warn that the population is put under general suspicion by open-street CCTV are denounced as 'protecting offenders'. Even though skepticism on the crime prevention effects of CCTV have meanwhile entered public debate, their arguments, pointing to the inappropriate and disproportionate nature of surveillance, die away unheard, as they sound weak in an age of "forensic battlefields", to borrow a term from Geraghty (1998). Once legislation and systems are in place, the role of data protection officers is, as Busch (2006) correctly notes, reduced to the supervisory though cosmetic prevention and persecution of abuse: they complain that cameras are recording in private homes, that surveillance data is stored longer than it should be, and that barriers between the observing parties in control rooms need to be created to prevent the sharing of data, but they have no means to fight the transformation of public space under the gaze of the camera.

Critics are usually assured by the advocates of open-street CCTV that it will be limited to crime hot spots and that nobody in Germany wants British big brother–style blanket surveillance. However, German police law does not in fact know the term *crime hot spot*. The relevant sections in the police law on open-street CCTV often refer to the definitions of 'dangerous places' where the police can authorize themselves stop-and-search powers, as described above. In Berlin, for example, the police designated more than thirty public places as 'dangerous' in the late 1990s—places at which they would theoretically be entitled to deploy cameras. Some police acts are even vaguer in their definition of areas which could be put under CCTV surveillance. The relevant section in the revised police act of the state of North Rhine Westphalia, for instance, defines areas by their "quality to encourage the commitment of criminal offences", a description which could, provided enough paranoid fantasy, apply to almost every location in a major city. But even in states where the relevant provisions are rather restrictive, the designation of 'crime hot spot' is put at the discretion of the police and can hardly be challenged, as crime figures are not publicly available for the microlevel. Thus, the assurance that CCTV surveillance in Germany will not expand beyond crime hot spots is either hollow or naive.

What actually limits the geographical expansion of open-street CCTV is a reluctant police force and the cost. Although key officials and all professional police organizations see open-street CCTV as a useful instrument in support of their work, influential forces within the police fear that the image of the police could be damaged by excessive use of surveillance as they would be unable to meet the rising expectations. Moreover, police culture is opposed to CCTV control room work, and fears that personnel could be replaced by cameras are important. In the field of protecting premises at risk the police have already experienced how CCTV has significantly contributed to the reduction in front-line policing. "The necessary police presence cannot be substituted or limited by technical surveillance" is the emphatic statement made by the Gewerkschaft der deutschen Polizei (Trade Union of the German Police), the largest professional organization, which represents the interests of 180,000 officers and employees of German police forces. Last but not least, open-street CCTV operated by well-trained police officers is an expensive affair, and even rather small systems are a significant load for the tight budgets of municipalities and German states. In Mannheim, the city government and the state of Baden-Württemberg were, on a shared basis, spending more than €600,000 for the first year of operation alone. In the state of Brandenburg the cost of the equipment and personnel for four pilot schemes was estimated to total €5.3 million euros at the beginning of a five-year test period. Moreover, police officers usually stress the "necessity to integrate CCTV into a comprehensive strategy" in order to make it work. Thus, the total cost covering the systems and control room personnel as well as those on the beat are even higher; and as those on the beat are usually there anyway, police managers see, in many cases, no need for cameras except when policing needs reassurance. Given this background it comes as no surprise that police departments in the cities of Cologne and Heidelberg dismissed the claims for cameras from local politicians, or that they were reluctantly submitting to the order of their governments as was the case in the state of Brandenburg. Instead the police seem to prefer flexible solutions such as mobile units which can be easily deployed at single events—for instance, the FIFA World Cup 2006, when many of the public viewing zones were designated CCTV areas; the annual October Festival in Munich; or other lesser crowd magnets such as Christmas fairs or carnival celebrations. Moreover, a recent trend is the integration of their command and control centers with camera networks of public transport and private premises, which provides flexible access while, in a sense, outsourcing the running costs. Transport corporations in Munich, Hannover, and Frankfurt am Main allow real-time access to their networks of hundreds of cameras. In Baden-Württemberg the police took an inventory of private cameras that monitor "areas and locations exposed to terrorist attacks" in order to achieve real-time access. However, the plan, backed by the conservative minister for the interior, was put on hold as the liberal coalition partner refused to support an amendment of the police act that could legalize the networking.

CONCLUSION

As we have seen, information privacy did not hinder the steady expansion of police powers in Germany. Their realization through actual surveillance was not limited by the right to informational self-determination but by the political economy of policing. Rather, the fair information principles derived from the constitutional right were considerably weakened. In particular, the flexible deployment of mobile units and the emerging "surveillance webs", to borrow a term from McCahill (2002), undermine careful consideration of whether a particular measure is necessary and appropriate and erode transparency and accountability. Thus, those who are watching are evading democratic control, even more so as the walls of these new *arcana imperii* are fortified through arguments that justify obscurity in the name of security.

But is the conception of privacy to be blamed for this? In Germany, it was the decreasing interest in the public value of privacy that enabled the nation's subjection to the rationale of selective crime control. Given the creeping character of how minor states of exception—declared by representatives of the executive branch of government—were expanded in time and space, the shifting of baselines enabled the easy colonization of the 'public interest' by powerful advocacy coalitions. In the case of public-area CCTV, this was made up of a corporatist network of internal security administrators, professional politicians, and members of the electronic industry and local businesses who were able to marginalize voices of critique.

The current threat to democratic polities that arise from such development is not only the possibility of a sudden inflation of minor emergencies into major ones but, even more, the selection and targeting of 'enemies' in the context of manifold minor states of exception beyond democratic control. The actual functions of surveillance may be opaque for all who are under the electronic gaze, but their panoptic power is targeting and silencing only a few (Norris and Armstrong 1999; Norris and McCahill 2006). But if the ability of these few to engage others is decided on an arbitrary basis, freedom—which, as Rosa Luxemburg once said, "is always the freedom of dissenters"—is transformed into a gift distributed by the powerful. Therefore, scholars of surveillance studies should think twice before denouncing the concept of privacy. They should avoid supporting those advocates of surveillance who attempt to devalue privacy of its social meaning.

NOTES

1. These are *intimacy, solitude, reserve,* and *anonymity*.
2. Since the 1970s these were engaged in campaigns of civil disobedience against the NATO policy to deploy cruise missiles in Western Germany, the building of nuclear energy plants, or other major projects such as the new western runway at Frankfurt airport. They were the grassroots from which *Die Grünen* (Green Party) emerged.
3. The census then took place on a modified legal basis and with revised questionnaires in 1987 but was still facing significant resistance and boycott.
4. If not stated otherwise the quotes are my own translations from *BVerfGE 65,1—Volkszählung* (Decision of the Federal Constutional Court 65,1), 15 December 1983; avaialble at http://www.servat.unibe.ch/law/dfr/bv065001.html.
5. These are the *Bundesamt für Verfassungsschutz* (BfV, Federal Office for the Protection of the Constitution), the *Bundesnachrichtendienst* (BND, Federal Intelligence Service) and the *Militärischer Abschirmdienst* (MAD, Military Counter-Intelligence Service). The 1990 Act revised the BfV Act and created acts on the BND and MAD, formerly unregulated by law.
6. This act included the authorization of electronic eavesdropping in private homes by an amendment of the Basic Law and is, therefore, seen as culmination of legislative action in the field of internal security in the 1990s.
7. Named after the then Federal Minister for the Interior, Otto Schily.
8. An interesting question in this context is the future quality of Cyberspace as the recent establishment of a "right to the integrity of personal computer systems", and perhaps the forthcoming decision on the Data Retention Act, points towards another direction.
9. Lange (2000) shows that policy making in the area of internal security in Germany is dominated by corporatist networks of representatives of the professional organizations of the police officers and their counterparts, often ex-policemen themselves, in the department for police of the Ministries for the Interior. Thus, the outcomes of the first draft bills presented by the executive branch of government to the parliaments do represent almost exclusively police interests.
10. "Zero Tolerance" and "Broken Windows" became prominent since the mid-1990s as indicated by the lead story on NY-style policing published in 1997 by *Der Spiegel*, Germany's largest newsmagazine. For a more detailed discussion of German policies against crime and disorder within the context of the "punitive neoliberal city" Herbert & Brown 2006 see among others:

Eick 1998; Ronneberger, Lanz and Jahn 1999; Simon 2001; Wehrheim 2002; Belina & Helms 2003; Eick, Sambale and Töpfer. 2007; Töpfer, Eick and Sambale 2007.

11. Besides these limited number of open street cameras thousands of police cameras monitor motorway traffic flows in Germany but these do currently neither collect personal data nor do they record footage. Moreover, at several locations fixed police cameras are installed that target public space but are not permanently in operation.
12. Which is the former *Bundesgrenzschutz* (Federal Border Police).
13. The *Zentralverband der deutschen Elektronik- und Elektroindustrie (ZVEI, Central Association of the German Electronic and Electro Industry)* is representing the interests of 1,600 companies, among others key players in the CCTV market such as Bosch. In charge of the lobby work for CCTV was their branch for security systems and the subbranch on CCTV. Meanwhile the ZVEI is collaborating with interest organizations of the private security industry BDWS and the BHE, another organization representing the interests of security technology companies, in developing standards for image data transfer between private CCTV surveillance and command and control centers of the police.

BIBLIOGRAPHY

Bäumler, Helmut. 2001. "Informationsverarbeitung im Polizei- und Strafverfahrensrecht." In *Handbuch des Polizeirechts*, 3rd ed., ed. Hans Lisken and Erhard Denninger. Munich: Beck, 735–911.

Belina, Bernd, and Gesa Helms. 2003. "Zero Tolerance for the Industrial Past and Other Threats: Policing and Urban Entrepreneurialism in Britain and Germany." *Urban Studies* 40(9): 1845–67.

Bennett, Colin J., and Charles D. Raab. 2006. *The Governance of Privacy: Policy Instruments in Global Perspective*. Cambridge, Mass.: MIT Press.

Bigo, Didier. 2006. "Security, Exception, Ban and Surveillance." In *Theorizing Surveillance: The Panopticon and Beyond*, ed. David Lyon. Cullompton, England: Willan, 46–68.

Bourdieu, Pierre. 1983. Ökonomisches Kapital—Kulturelles Kapital—Soziales Kapital. In Soziale Ungleichheiten, ed. Reinhard Kreckel. Soziale Welt 2: 183–98.

Brenneisen, Hartmut. 2003. "Videoüberwachung als Teil einer allgemeinen Sicherheitsstrategie." *Deutsches Polizeiblatt* 1: 9–13.

Busch, Heiner, Albrecht Funk, and Udo Kauss. 1985. *Die Polizei in der Bundesrepublik*. Frankfurt am Main: Campus Verlag.

Busch, Heiner. 1994. Gesetzesinflation und Parteienkartell. SPD- und Koalitionsentwürfe im Wahlkampf. *Bürgerrechte & Polizei/CILIP* 48: 31–38.

Busch, Heiner. 2006. "Hilfloser Datenschutz. Verrechtlichung, Individualisierung, Entpolitisierung." *Bürgerrechte & Polizei/CILIP* 85: 3–9.

Cohen, Stanley. 1985. *Visions of Social Control: Crime, Punishment, and Classification*. Cambridge: Polity Press.

Eick, Volker. 1998. Neue Sicherheitsstrukturen im "neuen" Berlin. Warehousing öffentlichen Raums und staatlicher Gewalt. *PROKLA. Zeitschrift für kritische Sozialwissenschaft* 28(1): 95–118.

Eick, Volker, Jens Sambale and Erik Töpfer eds. 2007. *Kontrollierte Urbanität. Zur Neoliberalisierung städtischer Sicherheitspolitik*, 1st ed. Bielefeld, Germany: Transcript-Verlag.

Ericson, Richard V., and Kevin D. Haggerty. 1997. *Policing the Risk Society*. Oxford: Clarendon Press.
Feeley, Malcolm, and Jonathan Simon. 1994. "Actuarial Justice: The Emerging New Criminal Law." In *The Futures of Criminology*, ed. David Nelken. London: Sage, 173–201.
Foucault, Michel. 1977. *Discipline and Punish: The Birth of the Prison*. New York: Vintage.
Geraghty, Tony. 1998. *The Irish War*. London: HarperCollins.
Heinrich, Stephan. 2007. *Innere Sicherheit und neue Informations- und Kommunikationstechnologien. Veränderungen des Politikfeldes zwischen institutionellen Faktoren, Akteursorientierungen und technologischen Entwicklungen*. 1st ed. Hamburger Studien zur Kriminologie und Kriminalpolitik 42. Münster, Germany: LIT Verlag.
Heitmeyer, Wilhelm, ed. 2008. *Deutsche Zustände. Folge 6*. Frankfurt am Main: Suhrkamp.
Herbert, Steve, and Elizabeth Brown. 2006. "Conceptions of Space and Crime in the Punitive Neoliberal Vity." *Antipode* 38(4): 755–77.
Hölscher, Michael. 2003. "Sicherheitsgefühl und Überwachung. Eine empirische Studie zu Einstellungen der Bürger zur Videoüberwachung und ihrer Erklärung." *Kriminologisches Journal* 35: 42–56.
Innenministerkonferenz. 2000. "Zur Veröffentlichung freigegebene Beschlüsse der 161. Sitzung der Ständigen Konferenz der Innenminister und -senatoren der Länder am 05. Mai 2000 in Düsseldorf." Accessed 31 March 2011 at http://www.im.nrw.de/inn/doks/imk0500.pdf.
Keller, Christoph. 2000. "Video-Überwachung: Ein Mittel der Kriminalprävention. Zum Stand der rechtlichen und taktischen Diskussion." *Kriminalistik* 54(3): 187–91.
Klocke, Gabriele, and Studiengruppe. 2001. "Das Hintertürchen des Nichtwissens. Was Regensburger BürgerInnen über die Videoüberwachung in ihrer Stadt wissen und denken." *Bürgerrechte & Polizei/CILIP* 69; http://www.cilip.de/ausgabe/69/video.htm.
Kunz, Thomas. 2005. *Der Sicherheitsdiskurs. Die innere Sicherheitspolitik und ihre Kritik*. Bielefeld, Germany: Transcript-Verlag. Kutscha, Martin. 2001. "Auf dem Weg zu einem Polizeistaat neuen Typs?" *Blätter für deutsche und internationale Politik* 46(2): 214–21.
Lange, Hans-Jürgen. 2000. "Innere Sicherheit als Netzwerk." In *Staat, Demokratie und Innere Sicherheit in Deutschland*, ed. Hans-Jürgen Lange. Opladen, Germany: Leske + Budrich, 235–55.
Leutheusser-Schnarrenberger, Sabine. 2008. "Auf dem Weg in den autoritären Staat." *Blätter für deutsche und internationale Politik* 53(1): 61–70.
Lyon, David. 2003. "Surveillance as Social Sorting: Computer Codes and Mobile Bodies." In *Surveillance as Social Sorting: Privacy, Risk and Digital Discrimination*, ed. David Lyon. London: Routledge, 13–30.
McCahill, Michael. 2002. *The Surveillance Web: The Rise of Visual Surveillance in an English City*. Cullompton, England: Willan.
Monahan, Torin. 2006. "Questioning Surveillance and Security." In *Surveillance and Security: Technological Politics and Power in Everyday Life*, ed. Torin Monahan. New York: Routledge, 1–23.
Müller, Rolf. 1997. "Pilotprojekt zur Videoüberwachung von Kriminalitätsschwerpunkten in der Leipziger Innenstadt." *Die Polizei* 88(3): 77–82.
Müller, Rolf. 2000. "Nochmals: Die Videoüberwachung von Kriminalitätsbrennpunkten in Leipzig. Eine Ergänzung zu den Abhandlungen in den Heften 3/97, S.77ff. und 4/98, S.114ff. ." *Die Polizei* 91(10): 285–92.

Narr, Wolf-Dieter. 1994. "Das 'System Innere Sicherheit'. Eine erstaunlich kontinuierliche Karriere." *Bürgerrechte & Polizei/CILIP* 48: 6–12.

Narr, Wolf-Dieter. 1998. "'Wir Bürger als Sicherheitsrisiko.' Rückblick und Ausblick." *Bürgerrechte & Polizei/CILIP* 60(2): 30–40; http://www.cilip.de/ausgabe/60/narr60.htm.

Normann, Lars. 2007. "Neueste sicherheitspolitische Reformergebnisse zur Terrorprävention." *Aus Politik und Zeitgeschichte* 12: 11–17.

Norris, Clive, and Gary Armstrong. 1999. *The Maximum Surveillance Society: The Rise of CCTV.* Oxford: Berg.

Norris, Clive, and Michael McCahill. 2006. "CCTV: Beyond Penal Modernism?" *British Journal of Criminology* 46(1): 97–118.

Regan, Priscilla M. 1995. *Legislating Privacy: Technology, Social Values, and Public Policy.* Chapel Hill: University of North Carolina Press.

Reuband, Karl-Heinz. 2001. "Videoüberwachung. Was die Bürger von der Überwachung halten." *Neue Kriminalpolitik* 13(2): 5–9.

Roggan, Fredrik. 2000. *Auf dem legalen Weg in einen Polizeistaat. Entwicklung des Rechts der Inneren Sicherheit.* Bonn: Pahl-Rugenstein.

Roll, Winfried. 2003. "Videoüberwachung im öffentlichen Raum zur Kriminalitätsbekämpfung." *Deutsches Polizeiblatt* 1: 2–8.

Ronneberger, Klaus, Stephan Lanz, and Walther Jahn. 1999. *Die Stadt als Beute.* Bonn: Dietz Verlag.

Roos, Jürgen. 2002. "Nichts geht mehr ohne Kamera. Nicht nur rechtliche Überlegungen zur Videoüberwachung." *Kriminalistik* 7: 464–69.

Sennett, Richard. 2000. *Verfall und Ende des öffentlichen Lebens. Die Tyrannei der Intimität.* 11th ed. Frankfurt am Main: Fischer-Taschenbuch-Verlag.

Simon, Titus. 2001. *Wem gehört der öffentliche Raum? Zum Umgang mit Armen und Randgruppen in Deutschlands Städten. Gesellschaftspolitische Entwicklungen, rechtliche Grundlagen und empirische Befunde.* Opladen, Germany: Leske + Budrich.

Stalder, Felix. 2002. "Privacy Is Not the Antidote to Surveillance." *Surveillance and Society* 1(1): 120–24; http://www.surveillance-and-society.org/articles1/opinion.pdf.

Töpfer, Eric. 2005. "Die Polizeiliche Videoüberwachung des öffentlichen Raums. Entwicklung und Perspektiven." *Datenschutz Nachrichten* 28(2): 5–9.

Töpfer, Eric, Volker Eick and Jens Sambale. 2007. "Business Improvement Districts—neues Instrument für Containment und Ausgrenzung? Erfahrungen aus Nordamerika und Großbritannien." *PROKLA. Zeitschrift für kritische Sozialwissenschaft* 37(4): 511–28.

Wehrheim, Jan. 2002. *Die überwachte Stadt. Sicherheit, Segregation und Ausgrenzung.* Stadt, Raum, Gesellschaft 17. Opladen, Germany: Leske + Budrich.

Weichert, Thilo. 1988. "Praxis und rechtliche Aspekte optischer Überwachungsmethoden. Zum Einsatz moderner Videotechnik." *DANA. Datenschutz-Nachrichten.* Special issue on video surveillance, 4–57.

———. 1998. "Audio- und Videoüberwachung im öffentlichen Raum." *Bürgerrechte & Polizei/CILIP* (2): 62–71.

Winkelmann, Arne, and Yorck Förster. 2007. *Gewahrsam: Räume der Überwachung.* Frankfurt am Main; Heidelberg: Deutsches Architekturmuseum; Kehrer.

8 The Pressure of the Practice
Swedish Public Surveillance in an Institutional Perspective

Elfar Loftsson

INTRODUCTION

The use of video surveillance by government, businesses, and individuals to watch activities in and around their space of responsibility has grown steadily in recent decades in modern societies and in many other parts of the world. The possibility to observe and record events at areas and places, for which they are responsible, attracts executives in private enterprise as well as in public administration, as it offers new and unique instruments for control. The advances in surveillance technology and its new applications have subsequently gained increasing interest from academics (Bennett 2011; Tække 2011; Monahan 2010). Still diverging in a disciplinary sense, and in many cases transdisciplinary in its approach, the predominant entrance into this field of study has long been Foucault's panopticonism (Wood 2003). Recently educational, sociological, and psychological theoretical perspectives have also been applied to the study of this phenomenon (Flaherty 1989). However, in this research area the literature is less extensive on the use of political approaches, particularly political institutionalism. In this chapter, public video surveillance in Sweden is investigated from an institutional perspective, aiming to scrutinise the surveillance phenomenon in Sweden by focusing on its values structure, rules, and practices.

An open debate on the partition between the individual and the collective in society is a natural part of the democratic public discussion. The question of what should be the proper role or responsibilities of the state and related authorities and what should be left to individual choice is, for many, a great controversy within political theory and in the main ideologies of Western democracies (Wood 2002). It is therefore no surprise that the question about new and improved techniques for surveillance and control, achievable for use by the state and other governing bodies, raises vociferous debates on what, where, and how such devices should be used and regulated. The phenomenon of public surveillance and control is by its very nature a controversial matter from a democratic perspective, but the rapid increase in capacity and number of such devices has amplified the debate.

In the Swedish political tradition an advanced and efficient public administration has a long standing, and has played an important role in the implementation of great public programmes, such as the structure and content of the welfare system, and more generally in the management of the public sector—something unique among Western democracies for its relative size.

Regardless of the ideological beliefs of successive Swedish governments, there has generally been a positive attitude towards large-scale technological surveillance in order to combat crime and antisocial behaviour, albeit with polite nods to the importance of privacy (Svenonius 2011: 140). In accordance with the Swedish administrative tradition, technological surveillance has been embedded in laws, instructions, and guidelines in order to promote uniform and legally trustful usage of the possibilities offered by the new surveillance technologies.

A commonly observed phenomenon is that societal matters during phases of rapid change require organisational changes in handling the dynamics of the issues concerned. It is therefore reasonable to assume that such changes tend to generate stress in the institutional structures intended to regulate them (Weaver and Rockman 1993). Thus, we also find that technically advanced surveillance systems in Sweden, as elsewhere, tend to cause concern as well as ideological controversies which are difficult for regulatory systems to understand and this lead to irregularity in practices. For these reasons, this chapter is devoted to an analysis of how technological surveillance in Sweden, in particular video surveillance, fits into and develops in the institutional structures formed to handle such issues. This aim is realised firstly by giving an overview of the concept of *institution* and, on that basis, constructing a coherent model of analysis; and secondly, by using the model to frame[1], describe, and analyse the public surveillance institutions in Sweden.

INSTITUTIONS AS ANALYTICAL CONSTRUCTS

Many scholars of new institutionalism, including March and Olsen (1989), refer to *norms* as a demarcation framing the concept of *institution*. However, norms seem to be synonymous with, or at least clearly related to, the concept of *values* as elaborated by, among others, Schwartz and Sagiv (1995). Even if of little help when trying to precisely define the concept, this conceptualisation marks an important improvement to the institutional theory in comparison with the 'old' one, as it makes clear that institutions do not exist in a normative vacuum (Lieberman 2003); rather, they are associated with, maintain, and reinforce culturally specific values and traditions. In most cases, though, the basic values of an institution cannot be found in its written statutes. Instead what we can expect to find is an institution with articulated objectives. But a closer look at such goals would surely reveal that they rest on more general values concerning something thought to be either good or bad—something worth promoting or preventing (Phillips et al. 2004). Regardless of the conceptual problems that may be related to the dimension of values/norms within this particular institutional pillar, many institutions do not exist just to maximise some 'good' goal. More often they are set up to strike a balance between different aims; however, if maximized, these aims may often contradict each other. This is clearly illustrated in the case of the surveillance institution, where two societal principles are at stake: personal privacy on the one hand, and *individual* as well as *collective* security on the other. In Sweden, for example, explicit goals for the public administration bodies responsible for surveillance express the intention that considerations of both security and individual privacy should be optimised when implementing surveillance.

Another conceptual core in institutional theory relates to *rules*. In several works on institutions, rules are simply referred to as "the rules of the game" (North 1991; Rothstein 1998). Whether this is to be understood so, that these authors regard the rule-system as constituting the entire structure of the institution, or that they also see other factors to be important, is not always clear (Droege and Johnson 2006). Most theorists within the realm of new institutionalism would agree that formal rules are among the basic elements of any institution. However, scholars stressing the importance of the rule system frequently do not only refer to formal rules but also to informal rules and constraints. Informal rules obviously rest on traditions and cultural values (Bohman 2000). Thus, including informal rules implies that the short-cut definition (rules of the game) reflects basically the same theoretical elements as expressed by March and Olsen, notwithstanding the difference in emphasis.

However, in both the afore-mentioned lines of thought, the normative- and the rule-centred, the material and operational aspect of institutions (including staff, organisational equipment, location, and handling of substantial output, etc.) are only given marginal attention, thereby limiting the analytical scope of what constitutes operational actions and behaviour. This lack of clear linkage to institutional practices has resulted in controversy regarding arguments about to what extent institutional theory is able to explain the outcomes of politics at all (Grofman 1997); and, subsequently, the debate on whether institutions matter has been rich (Przeworski 2004; Rothstein 1998; Weaver and Rockman 1993). Another critical view of current mainstream institutionalism is presented by Suddaby (2010), who finds the most promising development in institutional theory to be the efforts to analyse the role of language in institutional processes, referring to the importation of discourse theory and to the use of rhetorical approaches to study institutions.

Thus there seem to be good reasons to include those practice-related aspects, both in the theoretical modelling and in the analytical application when they concern institutional studies. A reasonable question here is whether the *practice* is not just that which the institutional construct should explain, and thus whether it could not be internalised into the analytical model (Lawrence and Suddaby 2006). But practice is here understood as denoting two things: (1) how internal order and organisation is formed, and (2) how the flow of output functions; thus, our suggested use of the practice concept does not limit the explanatory power of the model.

However, there are good reasons to expand a little on this concept of institutional practice. The two aspects of institutional practice that are given specific attention here focus firstly on the internal structure of an institution—that is, how it is organised, what professional skills are recruited and how internal change processes are managed—and secondly on the fact that most institutions are intended to deliver some kind of output and communication with other agencies in society (Scott 2005). The question of how this output production and the mutual contacts with the institutional environment are run may have great importance in determining the success or failure of the institution. Theoretically, scholars have handled these aspects differently. For example, Lindblom, among others, in attempts to modify traditional rational models, argued as early as the 1960s and '70s for an evolutionary approach to understanding institutions (Lindblom 1977). He introduced an incrementalist understanding of policy change where 'muddling through' situational issues was supposed to be a more realistic view of organisational behaviour than rational strategies for the fulfilment of fixed goals (Lindblom 1959). Simon's (1997) development of rational decision-making approaches, including among other things uncertainty and cost of information gathering, also added new insights to institutional theory. From an economic perspective, Nelson and Winter, in their seminal work *An Evolutionary Theory of Economic Change* (1982), stressed the importance of routines as crucial instruments in competition between firms, and for their success. What has been said thus far justifies the attempts to combine the three pillars of institutional theory into one model. Such an understanding of the concept of *institution* is illustrated in Figure 8.1

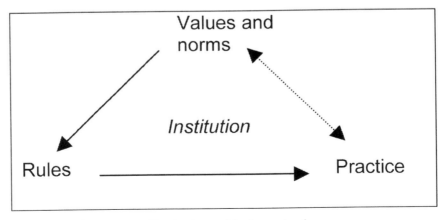

Figure 8.1 The concept of institution and its formative factors.

BRIEF REFLECTIONS ON THE MODEL

The arrows in Figure 8.1 indicate the rationale of the model. Thus, values and norms are supposed to determine the formation of the rule system, which in turn is expected to conduct the practice of the institution. *Practice*, on the other hand, is supposed to correspond to the values and norms.

Three further entrance points to this theoretical reasoning can be hinted at already at this stage: first, as the value-rules-practice pillars are seen as the fundaments of institutional structures, this implies that these pillars can be different in strength within an individual institution—that is, one institution can be predominantly value-driven, another rule-driven, and a third practice-driven. However, there are important structural differences between the three cornerstones of an institution. While the practices of institutions can be assumed to be sensitive to short-term changes, such as reorganisation and emergent demands from the operating environment, the value/norm component, on the other hand, is likely to be relatively resistant to change. Second, in the on-going scientific debate on development in Western democracies, the increasingly instrumental character of societal organisation has been pointed to (Bohman 2000; Sen 1993). In this view, the search for efficiency in the use of public resources leads to increasing scrutiny on the cost-effectiveness of organised activities. In this line of thought, it can be hypothetically stated that we expect the rationality and efficiency of practices to be given a great and growing importance in the development of institutions. Third, as the practices of an institution are likely to be occupied with the handling of short-term problems and demands, an increasing focus on these practical aspects may possibly lead to decreasing value awareness, ad hoc use of rules, and disruption of the routines of the institution—all contributing towards making a weakened and more unstable institution.

THE STRUCTURE OF THE SURVEILLANCE INSTITUTION

The theoretical discussion above has been elaborated in order to be relevant to the wide scope of *surveillance*, including both general and covert surveillance activities. In the following description, the focus is, when appropriate, narrowed to video surveillance in public spaces, leaving out the whole field of secret surveillance, the latter being primarily used by police to monitor suspected criminal activities.

As a means of power, more or less technically advanced surveillance is and always has been used in most societies, regardless of cultural and historical differences (Fonio 2011). Video surveillance can, for that reason, be studied within many traditional branches of analysis. However, as the phenomenon is undergoing substantial and fast changes, at least in technologically advanced societies, it challenges prevalent institutional theory—institutions are generally thought to be stabilising mechanisms in public life, and characterised by being resistant to rapid change (Hedlund 2007: 71).

Video camera use is an important part of surveillance, and the phenomenon has expanded rapidly in Sweden during the last two decades, both in terms of the number of cameras and in terms of regulations and administrative structures. According to a governmental investigation, the number of cameras that required formal permits to use such devices, and notifications about video cameras being installed, increased from about 5,000 in 1996 to some 18,000 in 2008 (Statens offentliga utredningar 2009: 68). Although no precise figures exist when it comes to the overall number of video cameras, the trend is clear. Thus, permits from the county boards grew by 45% in the period 2005–8 and, during the same time period, the number of cameras installed in schools increased by some 150% (Datainspektionen 2010).

Indeed, the development of this type of surveillance seems to be driven to a considerable degree by the increased possibilities of watching and recording events more effectively in public places. What we are witnessing, then, is a rapidly modernising security-related surveillance practice that is in tension with another, equally important, goal—that of guarding the individual privacy of citizens. So policy making has become a balancing act in which the question is how to make use of the potential of new and more efficient surveillance devices while at the same time minimising the intrusion into the personal privacy of citizens. In the following section, the analysis is in accordance with the modelling discussion above, thus distinguishing the *value prerequisites*, the *rule system*, and the *practices of the institution* in order to discover the consequences of how the institution is structured and operated.

THE VALUE DIMENSION

We have already discussed the idea of defining values as one of the basic cornerstones of institutions. On the formative level—that is, in the process in which the basic values and norms of the institution are manifest—the justification of the institution is stated. This process may include reference to general moral principles and/or to an on-going debate involving principles between different concerned actors. However, when it comes to empirical terms, and when the institution is established in an operational form, the basic values are transformed to more operational concepts in term of goals and ambitions. In most cases, general values are not referred to directly in constitutional documents of institutions, and when they are, it is mainly in vague terms (Droege and Johnson 2006). Nevertheless, almost all institutions have a well-defined set of goals which are supposed to guide the activities, development paths, and attitudes of the institution. We must, though, be clear that goals formulated in this way do not cover the complete scope of the value structure we have claimed to be part of the institution. The specific institutional goals are rather to be seen as embedded in the wider value structure of the society. But we can safely say that the explicit goals will reflect central aspects of the value structure.

A characteristic of the Swedish video surveillance institution during its formative period has been the duality in terms of goals; it combines goals related to individual privacy (*integritet*) and public security (*trygghet*, alternatively *säkerhet*; Björklund 2011). By no means uniquely Swedish features, these have been repeatedly stated in most authoritative documents concerning public surveillance; indeed, this division of goals constitutes the main focus in the debate on video surveillance (Marx 2002).

Despite different interpretations of those two value dimensions, they constitute truly central elements in any liberal democracy. Thus, the value basis of the surveillance institution is clearly framed by strong democratic principles. Notwithstanding the desirability of each of these values, the problem lies in their simultaneous maximisation; they can, when transformed into rules and practices, and implemented in policies, easily become contradictory. So, in Swedish legislation, as in the case of many other countries, the primary regulative ambition has been to shape an acceptable balance between their fulfilments. The incongruity that may occur between how much value is given to individual privacy versus that of security does not only take the placid form of preference ranking—that is, that one value can be enhanced when the other is kept stable. Rather, they seem to be negatively correlated, and the balance seems to be a zero-sum game so that an increased fulfilment of one is likely to lead to a reduction in the other. The zero-sum reasoning also indicates that the balance between core values is far more sensitive than if the values were not negatively related to each other. This also means that the surveillance institution, as we have defined and described it above, should more rightly be conceptualised in Sweden as a *security/privacy institution*—that is, there should be no exclusive promotion of either of the two core values but instead an optimal balance between them should be maintained in a way that is democratically satisfactory. However, here we will use the expression *surveillance institution*, as the focus is more on the limited surveillance activity taking place in public spaces than on the wide spectrum of the individual privacy-security issue area. Thus, we expect the underlying values to be empirically manifested in organisational goals.

In the preparatory documentation for the Public Video Surveillance Act (Lag 1998:150) these goals are clearly expressed by the parliamentary committee on justice: "When it concerns the fundamental motives for the current law proposal, the government stresses mainly the following in the law proposal: the existing law on video surveillance rests on the principle that information about cameras must be given and that a camera must not be used for the monitoring of spaces accessible to the public. If not, the interest of promoting the aim of surveillance outweighs the interest of privacy." (Justitieutskottets betänkande 1997/98:JUU14). Thus, the values of the surveillance institution are clear in the sense that they underline the need to grant both *security* and *privacy*. However, more precisely, the Public Video Surveillance Act focuses on where, when, and how public video surveillance can be utilised, and defines the restrictions on this, in order to create what is conceived as an acceptable balance between the two values, security and the privacy respectively.

THE RULE SYSTEM

The Swedish surveillance institution is primarily regulated by three acts of law: the Secret Video Surveillance Act, from 1995 (Lag 1995:1506); the Public Video Surveillance Act, from 1998 (Lag 1998:150); and the Personal Data Act (Lag 1998:204). In addition, the Secret Video Surveillance Act is complemented by the Act on the Prevention of Certain Particularly Serious Offenses (Lag 2007:979). A number of other laws and provisions link to these acts of law in specific paragraphs.

Provisions for public video surveillance in Sweden date back to the The TV-Surveillance Act, from 1977 (Lag 1977:20) and are now codified in the Public Video Surveillance Act[2]—the law of primary interest here. This law consists of introductory provisions and detailed sections concerning, among other things, disclosure requirements, permits, notifications, storage of materials, procedures and decisions concerning permits and notifications, and provisions on supervision and confidentiality. Further, the legal limits are also clarified by stating the law to be applicable in places to which the public has access, as distinct from those places to which the public does not (Lag 1998:150, 3§).

The structure of the act embodies the main intentions of the legislature, and thus deserves further comment. First, the law is effective for optical as well as acoustic observation and recording. The introductory provisions, in addition to demarcating the borders with other law and presenting initial definitions, make it clear that the law is aimed at creating a stable balance between the two core values of the institution: *privacy* and *security*:

> §1. In this law provisions are stated about the use of surveillance devices (public video surveillance). Public video surveillance must be executed with proper respect for the individual's individual privacy. . . .
>
> §6. Permission for public video surveillance shall be given if the interest for such surveillance outweighs the individual interest of not being observed by the camera. (Lag (1998:150).

The rather general and restrictive formulations of the initial paragraphs of the law are in some cases given a more precise form, and also open up to account for exceptions from the restrictions. The central concepts—security and privacy—are given only a vague definition in the law. Regarding the security aspect, it is said that when the benefit of installing video surveillance is assessed, special attention shall be given to whether the surveillance is needed to prevent crime, avoid accidents, or similar purposes. Concerning the privacy aspect, it is stated that when assessing the individual's interest in not being monitored, special attention shall be given to how the monitoring is operated, and what area is to be monitored (§6). Further, in the provisions concerning permits it is for example stated that public surveillance without a permit is allowed during one month. If an application has been submitted to the county administrative board (one in each of the twenty-one administrative counties) during this period, it can be done even up to the point that the board has dealt with the application (§10). Given the possibility of appealing to the attorney general in the event the application is refused, the time frame for usage of video cameras without a permit can be substantial.

The disclosure requirements state that the use of public surveillance cameras shall be indicated clearly by signs or in some other suitable way, and that if the equipment allows sound recording, this shall be also indicated separately. Exceptions from both permit and disclosure requirements are many, and relate among other things to traffic watching, protection of vital public installations, and places assessed by the police at serious risk of crime (§§7–9).

In order to make it easier for banks, post offices, and stores to utilise surveillance devices, the law states that in many cases it is sufficient for these parties to make note of the surveillance to the concerned county administrative board. A restriction here is that the surveillance is solely aimed at crime prevention and detection and that the camera is mounted and fitted with fixed optics (§11). In accordance with these provisions, stores are allowed to use cameras at entrances and checkouts after giving notification. If other areas of the store are to be watched, permits from the county administrative board are needed. Surveillance in areas and buildings other than banks, post offices, and stores requires a permit.

When it comes to the storage of recorded material, the law states that such material shall be handled so that misuse is prevented, and that persons other than those needed to execute the surveillance shall not have access to such recorded pictures or sound (§13). Further, video and sound recordings from such surveillance may be stored for no more than one month, unless the relevant county administrative board gives permission for longer storage (§14). In additional provisions from 2004, exceptions to these requirements were incorporated regarding cases (1) when access to the material was considered to be of importance for investigation of criminal activity, (2) when the material had been delivered to the court in proceedings on responsibility for crime, and (3) for material stored by video cameras used for traffic taxation within Stockholm city (Lag 2004:633).

Swedish legislation is in general terms consistent in the sense that all new acts of law are passed only after such revision of existing law that there should not be contradictions. Thus, the Public Video Surveillance Act has had eleven additional provisions since it came into effect. These changes have only marginally extended the scope of the regulations, and have mainly reflected changes in the other laws referred to in the act. However, there are still some confusions in the relationship between the Public Video Surveillance Act and the Personal Data Act, where the latter is generally, in the case of incongruence, subordinate to other legislation but at the same time takes precedence in the parts concerning video surveillance, which are not stated in the Public Video Surveillance Act. A solution to this problem has been suggested in the governmental committee's proposal for a new video surveillance law (Statens offentliga utredningar 2009).

The Personal Data Act was created to meet the requirements of the data directive EC 95/46 of the European Union (EU). The law is applicable to surveillance in places to which the public does not have admission—for example, workplaces, private houses, and schools. When it comes to the kind of surveillance we focus on here—namely, video surveillance—this law is only applicable when personal identification is possible from recorded or processed materials.

THE PRACTICE

As discussed earlier in this chapter, *practice* as a basic institutional aspect is here conceived of as two different processes. First, it concerns the interior of the institution, which includes among other things the dimensioning, structure, and internal rules of the organisational body. Second, *practice* is about the interaction with the societal environment. This includes both handling of demands from the environment and the institution's deliveries to society, though in many cases there is a close relationship between these two aspects of practice. An organisational change within an institution is likely to affect its services to society and, similarly, changes in the institutional environment are likely to force internal changes in the institution. An important factor in this respect is the rapid and on-going development in the performance of surveillance devices.

Technological Development

An important driver of pressure on the video surveillance institution is the development of surveillance technology. Since the first use of television surveillance in the mid-1960s, the technological development has been revolutionary. At that time, surveillance was based on simple closed-circuit television. The next important step in the development was the introduction of videocassette recorders, based on analog technology, which allowed the recording of substantial amounts of data. Passing stages of microchip-equipped coupled-device cameras and digital multiplexing, offering new flexibility and improved recording quality as well as saving on videotapes, the technology developed further. The introduction of digital technology-based systems marked a new epoch in the development of video surveillance (Introna and Wood 2004). Instead of using recorded tapes, an immense amount of data could now be stored on hard drives. In addition, the picture and sound quality improved dramatically. Still, the latest development in surveillance-applicable technology—Internet protocol technology—seems to be only in its premature stage. The streaming of data produced by extremely flexible devices such as mobile phones remotely controlled through the Internet and situated at any distance from the surveillance point imply practices that increasingly fall outside the competence of existing governmental regulations and control (Roberts 2005).

There seems to be an agreement among experts about the tendency for practically all current technological development in the sphere of video surveillance to take place in the field of digital technology (Statens offentliga utredningar 2009: 80ff), whilst analog technology, even when transferred digitally, is successively declining. Still, purely digital systems seem to be more expensive when installed, mainly due to the more expensive digital cameras. However, as the quality of digital systems is much better in terms of picture quality, face recognition programs, and the like, this technology makes alternatives more or less redundant. Another aspect of these technological innovations concerns the additional benefits to both criminals and law enforcement authorities. Advanced mobile devices for surveillance are equally available to both criminal and law enforcement actors. This is likely to lead to a technology race between the parties, and to protect privacy in such a climate may thus become a gradually more difficult task.

Technological improvements in surveillance equipment are certainly an international phenomenon, but they have consequences at the national level as well, and also for the general administrative structure of the Swedish institution. It seems obvious that the increase in surveillance points, together with the radical technological improvement of the equipment, presents a challenge to the capacity of public administrative bodies responsible for the video surveillance institution. In that context it is worth mentioning that the responsible authoritative bodies in Sweden have not expanded in terms of staff number in proportion to the increase in video surveillance. And the two main bodies—the county administrative boards and the national Data Inspection Board—have only to a limited degree been able to add professional skills appropriate to the changing situation in this field (Data Inspection Board 2008; County Administrative Board of Stockholm 2008). On the other hand, the internal organisation of both these bodies has undergone changes, primarily due to the implementation of EU initiatives.

In most cases of centrally formed policies intended at nationwide implementation, different public administrative bodies bear the burden of realising the ideas and ambitions expressed in norms and rules behind these policies. This means that international agreements and national legal acts, as well as different instructions and local provisions, are to be applied by many different bodies of public administration and by different authorities. This also means that institutional implementation is likely to take different forms in different local settings.

There are several authoritative bodies involved in the institutional practice of surveillance—primarily the county administrative boards, the Data Inspection Board, the police, the attorney general, and the concerned municipal authorities. However, the county administrative boards and the Data Inspection Board are the most important in terms of organising and handling the flow of surveillance cases. Their central positions rest on their responsibility for the execution of the two main acts of law discussed above that regulate the surveillance institution. The other bodies have more limited and specific formalised roles related to the surveillance institution.

The county administrative boards are the responsible authorities when it comes to execution of, permits for, and supervision of video surveillance in public places. The practice here is mainly contained in the process of permits and in supervising and control. The role of the county administrative boards in the practice of video surveillance is by nature most exigent in Stockholm and in other of the main cities. As the possibility for certain actors to install cameras has been extended, with the only administrative requirement being to deliver notification to the county administrative boards, the most important task of the boards has become to handle permit applications; requests from the public transport companies comprise some 80% or more of those (County Administrative Board of Stockholm 2008).

The Data Inspection Board is involved in video surveillance in places to which the public has access only in cases when the Personal Data Act is relevant. In general, that means situations in which personal identity can be uncovered. According to the inspectorate, there were only seventeen such cases in which it became involved in the period 2005–8 (Data Inspection Board 2008). However, the Data Inspection Board plays an indirect role. First, the board receives information about decisions taken by the regional county administrative boards and in some cases gives its views on them. Second, the board groups its own surveillance cases into categories according to their character and function. Thus, schools, residential issues, and the like constitute different types of problems. In so doing, the inspectorate produces policies that are communicated to other bodies involved in surveillance control and therefore have the possibility to influence these communicating partners.

As indicated above, the implementation of laws and policies normally involves different administrative bodies. Video surveillance is certainly no exception in this respect. Notable differences in the surveillance practices of different local authorities are underlined in two major evaluations of the Public Video Surveillance Act that have been undertaken since its adoption in 1998 (Statens offentliga utredningar 2002: 194ff; Statens offentliga utredningar 2009: 178). This lack of coordination is confirmed by our own interview with the vice director of the unit for permissions and supervision at the county board (County Administrative Board of Stockholm 2008). In the most recent of the governmental evaluations, one of the proposals is that the Data Inspection Board should have responsibility for all control under the Public Video Surveillance Act. One main argument for this is the diversity of controlling practices on the part of the different county administrative boards. The investigators stated that the county boards had handled the surveillance issues in a satisfactory way, but pointed to two problems—namely, the lack of uniformity in rule applications and in control (Statens offentliga utredningar 2009: 180).

An important responsibility of the attorney general is to inspect the county administrative boards and to supervise their decisions and administrative routines. In none of the six inspections undertaken between 2002 and 2008 did the attorney general make any critical reviews concerning the surveillance practices of the county administrative boards. The field seems to have had a limited priority in the overall inspections, and it is only given brief attention in the inspection reports (Welsh and Farrington 2007, Justitiekanslern 2012). In relation to the regulation of video surveillance in public spaces, the attorney general's office describes its role in the regulatory system for video surveillance in places to which the public has access: "The attorney general has the right to appeal decisions by the county boards concerning video surveillance (Lag. 1998:150). The reason for this is to protect the public interest that a reasonable balance is held between the interest to impeach crime and the individual's interest of respect for her privacy. The administrative courts and ultimately the Swedish supreme administrative court decide how this balance shall be practiced in different situations" (Justitiekanslern 2011). The office of the attorney general, in its role as the appealing authority, gives an input as to how rules should be interpreted and practices made routine. In that sense, the authority contributes to the formation of the practices of video surveillance.

The municipalities certainly have a formal say in the practice of video surveillance, but with the exception of the main city municipalities, their role tends to be quite limited and diverse. Usually they tend to approve plans and activities proposed by the county administrative boards, the Data Inspection Board, and the Stockholm Police (SKL 2010). However, the municipalities have become more active in taking and supporting initiatives relating to video surveillance within their borders. Generally, such initiatives are related to situational problems—for example, local criminality in certain streets or residential areas and problems in and around schools (Fjellman Jaderup 2012; Svensson 2010). In many municipalities policy documents have been adopted and, during 2008 and 2009, the Swedish Association of Local Authorities and Regions invited the county boards to a series of conferences on the preventive effects of video surveillance (Brå and SKL 2009). It is clear that the municipalities vary greatly in terms of their needs and capacities, and therefore in their approach to video surveillance policies and routines. The Data Inspection Board noted, after an extensive evaluation of video surveillance in schools across the country, that in many cases cameras were installed and run in contravention of the Personal Data Law. The board included seven schools in this evaluation and found that in only one case was the use of cameras in accordance with the law (Datainspektion 2010).

Although in some respects peripheral to the policy-making and administrative handling of video surveillance, the police undoubtedly constitute an important actor in relation to this institution. And on matters of secret video surveillance, the police have implementing authority. Most of the security-related provisions in the central acts of law concerning the video surveillance relate to crime prevention and investigation, for which the police constitute the central operational body. And at most of the different steps for the installation of video surveillance cameras, advice or instructions from the police are requested. Thus, the police take part in the planning of video surveillance and assist surveillance users in different ways to create efficient and legal surveillance systems; for example, the police, in cooperation with the Swedish Trade Federation, the Swedish Bankers' Association, and the Swedish National Laboratory of Forensic Science, recently published detailed guidelines for surveillance camera users (Statens kriminaltekniska laboratorium 2005).

The police generally promote video surveillance as a means of *combating* crime, which is likely to be an important factor in gaining wider acceptance of this kind of surveillance (SLTF 2006: 1). However, the police operates primarily under the provisions of the Police Law, and consequently the rule system of the video surveillance institution, in its public part, is but one of many regulations created or modified in order to coordinate with the police functions.

Permits and Control

According to the Public Video Surveillance Act, video surveillance users have to apply to the county administrative boards for permits to install surveillance devices. These regional authorities are also responsible for the supervision of the regulations. In Stockholm, it is the business division within the county administrative board that handles the video surveillance issues, and here permit issues dominate. According to the diary register, permit issues comprise over 90% of work related to surveillance (County Administrative Board of Stockholm 2008). Permit applications are made on a pre-printed form, which requires, among other things, the reason for surveillance. The county administrative board then passes the application to the concerned municipality's administrative unit. The county board or the municipality frequently requires additional information from the applicant. However, in most cases, permits for video surveillance are approved by the county board, even though refusals sometimes occur, as in the case when the Stockholm local transportation agency filed an application to install video cameras at the entrances to bus and underground stops (County Administrative Board of Stockholm 2008).

In public locations such as banks, post offices, and shops where the public has access, an authoritative permit is not needed, but a notification to the county administrative board should be made. In other open spaces permissions are required. Those who install surveillance equipment are obliged to make people aware of it, which is usually done by posting signs within the monitored area. The county administrative boards are obliged to assess the necessity of the surveillance against personal integrity interests in the approval process. In their role as the supervising authorities, the county boards are also responsible for surveillance installations being done in accordance with given reports and permits (Lag 1998:150, §18–24).

Thus, the regional county administrative boards are responsible for the implementation of the Public Video Surveillance Act in the sense that they investigate the need and issue permits for video surveillance and control the correct enforcement of the permits. In the governmental report delivered to the ministry of justice in 2001 the investigators state, "The need for increased control is general in the entire field of video surveillance, but is particularly clear concerning the permission-free surveillance in banks, post offices and shops" (Statens offentliga utredningar 2002: 183).

The other main actor regarding the practice of the surveillance is, as discussed above, the Data Inspection Board, which is the liable authority when video surveillance affects the Personal Data Act, which is the responsibility of the inspectorate. This may concern video surveillance and recording where identification of individuals is possible. Here, though, there is an important exception: the inspectorate is not responsible for video surveillance in public spaces to which the public has access. Therefore, the inspectorate becomes involved only if video cameras are to be installed in a new category of places. Once the permits for a new surveillance category have been decided upon, the inspectorate only formally becomes involved by receiving the regular reports from the county administrative boards (County Administrative Board of Stockholm 2008).

Characteristic of Swedish institutional practice is very limited contact between the main actors. This was confirmed by an interviewee at one of the county boards, who described contact between the other county boards, as well as with other concerned authorities, as "very limited" (County Administrative Board of Stockholm 2008). At the same time, this interview, as well as the documents on file from this administration, show clearly that surveillance practice is, to a high degree, managed in routines, which means that working tasks have been categorised and the procedures for handling them have been made uniform. However, in its report and proposals to the government, the special committee set up to revise the Video Surveillance Act explicitly mentions the lack of coordination between the different concerned parties responsible for the regulation of surveillance practice (Statens offentliga utredningar 2009: 18). The committee considers increased coordination between different county administrations, as well as other bodies responsible for video surveillance, as an important improvement that needs to be made.

We can already see at this stage that, despite internal organisational changes, both the value- and rule-related components of the institutional set-up are rather stable—that is, in the terms in which they are formulated and expressed in authoritative documents. Indeed, changes in the law have, to a certain extent, opened up the wider use of video surveillance. The most remarkable changes we have witnessed relate to practice. The number of users and the number of cases handled by the concerned authorities have increased dramatically. Parallel to this, the technical improvement and increased capacity of the devices has been substantial. Thus, the changes we see to the video surveillance institution seem to have taken place primarily relate to the extent of its practice, both *internally* (such as in the organisation and allocation of responsibilities) and even more *externally* in terms of the implementation.

DISCUSSION

The foregoing analysis has described video surveillance in Sweden from the perspective of institutional theory. This means that the main focus has been on the value-reflecting goals and the regulatory structures guiding the implementation. In this section, the main results are discussed and the relevant conclusions are presented and related to our model of analysis. Two central questions remain to be discussed further—namely, what is the character of the institution we are left with and what are the wider societal consequences that we can expect from its development?

An Institutional Transformation

The regulations in Sweden take into account that there must be limits to how far central value concepts can be protected. Individual rights must be weighed against collective needs. In this case there are many societal values that must be taken into account, such as security, vital economic interests, and common welfare, to mention but a few. The most vital, however, seems to be the security aspect, and to assure people that increased video surveillance also means increased safety in public places. In the legislation regulating this field, we have seen that this is the chief concern. The legislation refers to keeping a balance between the two central values of *security* and *privacy*, although these are only vaguely defined. Together with the many exceptions stated to the general intention of the law, the use of such vague phrases and concepts as "permit for public video surveillance shall be issued if the interest for such surveillance outweighs the individual interest of not being observed by the camera" is frequent (Lag 1998:150, §6).

When we turn our attention to our model of analysis, and the main pillars of the surveillance institution, we notice that the Swedish surveillance institution is still resting on its basic value system, on the comprehensiveness of the rules system, and the practices to be steered by values and rules. But the institutional structure has also undergone substantial change that that well may lead to a severe weakening of the institution itself.

From Value Balance to Value Hierarchy

When examining the central intentions of laws regulating video surveillance in Sweden, we have seen that the relationship between the two central value aspects—enhancement of *security* and protection of *privacy*—is thought of in terms of achieving a balance between them. During the period from the early 1990s up to the present day, these values have been legally held to be of equal importance. However, as practice has developed, through, among other things, rapid technological development and an increased use of video surveillance, this balance seems to be in a process of change. It is obvious that improved technology and the growing numbers of cameras make it more difficult for a person to hide or be anonymous in public spaces. No parallel development has taken place to strengthen the protection of privacy. Authoritative statements concerning video surveillance increasingly stress the need for improved devices to combat criminality, but many fewer initiatives and ideas are devoted to privacy aspects. The zero-sum character of the contest between these two core values, as discussed in this chapter's theoretical section, illustrates the sensitivity of the value balance. The development of video surveillance that has been described raises questions about whether it is realistic to conceive of the two value-related interests in terms of a balance. Indeed, it seems justified to describe the changes taking place here as a transition from a balance between *security* and *privacy*, to a duality of these values in a ranking order, where security is given the higher priority. This is also reflected in a tendency to discuss the two desired aims separately and not in the same context, which among others is illustrated by the numerous motions tabled in the Riksdag by members of parliament. (e.g., Motion 2011/12:Ju209, Motion 2011/12:M852).

FRUSTRATION IN THE REGULATORY SYSTEM

Our inventory of the Swedish surveillance institution has shown that considerable and fast development has taken place in the practice of video surveillance. According to our theoretical discussion above, we would expect this to result in changes to the basic regulatory structure of the institution. Among other things, we would expect rules to become irrelevant or applied to unforeseen situations when new practice matters occur. In fact, we can see indications of such a development. In the two major investigations of the Public Video Surveillance Act we have referred to above, a number of legal problems were highlighted. The second of these investigations, "A New Video Surveillance Law" (Statens offentliga utredningar 2009), had directives to propose necessary changes to the relevant laws. The investigators duly proposed five important changes to the regulatory sector, including, among other things:

a) That all provisions about public video surveillance, with exception of private video surveillance, are to be gathered in one law, 'The Law on Video Surveillance'.
b) Privacy protection is to be strengthened by extending earlier special provisions about video surveillance in public places, to all video surveillance under the law.
c) Introduction of payment for damages, meaning that the parties executing such surveillance illicitly are to pay compensation to the parties being observed, for the infringements of privacy caused by the processing of pictures and sound recordings.
d) Video surveillance, including the right to record pictures in shops after notification. The storage period for materials recorded through public video surveillance, at places to which the public has attendance, should be prolonged from one to three months at most.
e) The Data Inspection Board should have the central responsibility for the application of the law. Here was included evaluation of the law application, following up on the supervision, advice, and support to the county boards and publishing of general guidance (Statens offentliga utredningar 2009: 13–19).

However, despite commitment from successive governments in power during the process of updating and restating the law in this area, there have so far been no such governmental bills proposed to the parliament. There could be many reasons behind this hesitation. First, on-going regulatory reform preparations are taking place at the EU level that may be causing Swedish politicians to suspend their own initiatives. Second, a judicial reformation of the rapidly increasing and diversifying surveillance field implies a vast political undertaking, and even a thorough redressing such as this would most likely be in immediate need of changes by the time of completion. Third, the recent legislation, under the umbrella of the so-called FRA law, has drawn attention away from the specifics of video surveillance. The initiative to extend the duties of the National Defence Radio Establishment (Försvarets radioanstalt, FRA) to include signal intelligence on information passing on cables over the Swedish border was taken by the social democratic government in the mid-1990s and has since been followed up by subsequent governments. As the law reforms concerned, among other things, a large part of Internet traffic and personal information, these reforms have lead to widespread public protests (Lag 2008:717).

Finally, the generally strong public support for video cameras in public places does not promise much political gain from politicising the issue. There seems to be little doubt about the positive attitude among the general public towards video surveillance in Sweden, though people seem also to overestimate both the safety benefits in supervised areas and the crime reduction gained by such surveillance. The facts are, however, when looking at the totals, that the steadily growing use of video surveillance cameras does not correlate positively with general crime reduction. Evaluations of camera use show only a marginal reduction in crime in areas under surveillance. Also, when it concerns people's general perception of safety it does not show corresponding improvement in relation to the increased use of this kind of surveillance.

However, there *are* some notable effects of the surveillance undertakings. The public debate on privacy and security has definitely become both more sophisticated and more intense, although the debate primarily appears as technical and rational in character. So there seem to be many reasons for the hesitation in introducing a new law in the field of video surveillance. But even if the proposed changes have not been manifested in new legislation, they already have effect in the sense that all propositions in the parliament regarding video camera usage have been rejected with reference to the necessity not to anticipate the on-going regulative preparations (Justitieutskottets betänkande 2010/11:JuU9).

The public popularity of camera monitoring obviously encourages party initiatives to accept parliamentary initiatives and even bring about changes in the legislation. For instance, ten out of twelve initiatives and motions to the Riksdag in the period June 2010 to October 2011 were positive in one or another way in terms of increasing video camera use. Of those ten positive motions, the moderates proposed four, and the liberals, the Centre Party and the Sweden Democrats two initiatives each, while the two negative motions were proposed jointly by the Social Democrats, the Left Party, and the Green Party.[3]

I have argued above that there are clear indications of practice stress in the Swedish video surveillance institution. This situation becomes clear when we notice that all the main changes taking place within the institution concern the increase in scope, extent, and quality of video surveillance. At the same time, the regulatory system has been only marginally changed, and then has opened up the wider use of video surveillance. Also, in the field of values and norms behind the institutional set-up, changes seem to have taken place in the direction of giving a higher priority to the security aspects judged to be fulfilled by the surveillance. A question arises here—namely, what effects can we expect from this development on the function of the institution of surveillance?

Obviously the changes we have observed may bring about severe effects on the functioning of the institutional framework within which public video

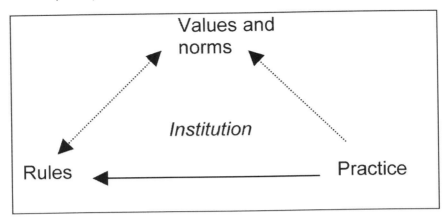

Figure 8.2 The structure of a practice-driven institution.

surveillance is handled. First, as was highlighted above, public video surveillance is in a phase of institutional practice stress, which has consequences also for the other main cornerstones of the institutional structure. We have described this step in the development as 'regulatory frustration' in the sense that the input overload requires ad hoc solutions where existing rules have to be reinterpreted or new rules need to be established. In addition, the value basis of the institution may become less relevant as the rules, in their application, do not correspond to values and norms but instead to the current needs of handling day-to-day issues. Figure 8.2 indicates the weakened relationship between the value structure and the other basic components of the system, and the increased practice influence within the institution of surveillance:

In contrast to Figure 8.1, in which the values and norms were supposed to determine the formation of the rule system and guide practices in harmony with the values, the institution described in Figure 8.2 is characterised by practices that inform the actual rule system. These practices then fall out of step with the supposed basic values of the system and both rules and practices get a diffuse relation to these values.

CLOSING REMARKS

The rapidly changing prerequisites for public surveillance are clearly manifested in the recent developments in public surveillance in Sweden. While the reformation of regulatory and administrative instruments is getting more and more complicated and time-consuming, the technological improvements of surveillance devices develop swiftly. The general principle of Swedish public video surveillance has been to restrict the information gathered by surveillance activities. It seems likely that this principle may be difficult to maintain in the future. Those who want information of this kind, or who want to fabricate information, and are reasonably skilled in the use of the new technology are likely to get what they want.

Another, and maybe more severe, aspect of the technological practice drive of the surveillance institution relates to the general political life as being either *proactive* or *reactive*. Sweden has a tradition of politics being proactive. During the long period of social democratic rule, political ideals were formulated and championed. This led to the launching of programmes and policies aimed at shaping a societal future in which the welfare society was the core concept. Over the course of recent decades, this political approach seems to have progressively turned into a more *reactive* politics. Instead of fulfilling societal ideals, politics has to put effort into adjusting and legitimating on-going practices in society outside of political life. This tendency seems to be well demonstrated in the development of the institution of video surveillance. The ideals of a society in which privacy has traditionally been seen as a symbol of individual freedom to be combined with collective security have gradually become eroded in the face of demands for quick responses to security problems and advanced criminality. If these tendencies become a more general trend, the political preconditions would be dramatically changed as the institutions—the main stabilising structures of political development—lose this vital function.

NOTES

1. *Framing*, as used here, refers to the associations and cognitive effects that theoretical concepts have due to their inherent meaning. Nelson et al. give the following definition: "Framing is the process by which a communication source, such as a news organization, defines and constructs a political issue or public controversy" (1997: 221).
2. A new law is under preparation in this field. A governmental committee has delivered a final report proposing both coordination among the parties concerned in the implementation of the law and changes concerning its practice. Still, though, the government has not proposed anything new on the issue. See Statens offentliga utredningar 2009.
3. Moderates: Motion 2011/12:Ju264; Motion 2011/12:M852; Motion 2011/12:M24; Motion 2010/11:Ju279. Liberals: Motion 2011/12:Ju209; Fråga för skriftligt svar 2010/11:448. Centre Party: Motion 2011/12:Ju280; Motion 2010/11:Ju304. Social Democrats: Motion 2011/12:Ju261; Motion 2011/12:SD66. Social Democrats, left Party, and Green Party: Motion 2010/11:Ju8; Motion 2010/11:-S96005).

REFERENCES

Bibliography

Bennett, Colin J. 2011. "In Defence of Privacy: The Concept and the Regime." *Surveillance and Society* 8(4): 485–96.

Björklund, Fredrika. 2011. "Pure Flour in Your Bag: Governmental Rationalities of Camera Surveillance in Sweden." *Information Polity* 16(4): 355–68.

Bohman, James. 2000. *Public Deliberation: Pluralism, Complexity and Democracy*. Cambridge, Mass.: MIT Press.

Droege, Scott, and Nancy Brown Johnson. 2006. "Broken Rules and Constrained Confusion: Toward a Theory of Meso-Institutions." *Management and Organisation Review* 3(1): 81–104.

Fjellman Jaderup, Elin 2012. "Reepalu kör sitt eget race". *Sydsvenskan*. March 25, 2012.

Flaherty, David H. 1989. *Protecting Privacy in Surveillance Societies*. Chapel Hill: University of North Carolina Press.

Fonio, Chiara. 2011. "Surveillance under Mussolini's Regime." *Surveillance and Society* 9(1–2): 80–92.

Hedlund, Stefan. 2007. *Instutionell teori: Ekonomiska aktörer, spelregler och samhällsnormer*. Lund: Studentlitteratur.

Introna, Lucas D., and David Wood. 2004. "Picturing Algorithmic Surveillance: The Politics of Facial Recognition Systems." *Surveillance and Society* 2(2–3): 177–98.

Lawrence, Thomas B., and Roy Suddaby. 2006. "Institutions and Institutional Work." In *Handbook of Organization Studies*, 2nd ed., ed. Stewart R. Clegg, Cynthia Hardy, Thomas B. Lawrence, and Walter R. Nord. London: Sage, 215–54.

Lieberman, Robert C. 2003. "Ideas, Institutions and Political Order: Explaining Political Change." *American Political Science Review* 96(4): 698–712.

Lindblom, Charles. 1959. "The Science of Muddling Through." *Public Administration Review* 19: 79–88.

Lindblom, Charles. 1977. *Politics and Markets: The World's Political-Economic Systems*. New York: Basic Books.

March, J. G, and J. P. Olsen. 1989. *Rediscovering Institutions: The Organizational Basis of Politics*. New York: Free Press.

Marx, Gary T. 2002. "What's New about the "New Surveillance"? Classifying for Change and Continuity." *Surveillance and Society* 1(1): 9–29.

Monahan, Torin. 2010. *Surveillance in the Time of Insecurity*. New Brunswick, N.J.: Rutgers University Press.

Nelson, Richard R., and Sidney G. Winter. 1982. *An Evolutionary Theory of Economic Change*. Cambridge, Mass.: Harvard University Press.

Nelson, T. E., Z. M. Oxley, and R. A. Clawson. 1997. "Toward a Psychology of Framing Effects." *Political Behavior* 19(3): 221–46.

North, Douglass. 1991. "Institutions." *Journal of Economic Perspectives* 5(1): 97–112.

Peters, B. Guy. 2001. *Institutional Theory in Political Science: The "New Instututionalism."* London: Continuum.

Nelson, Phillips, Lawrence, Thomas B. and Hardy, Cynthia. 2004. "Discourse and Institutions." *Academy of Management Review* 29(4):635–632.

Przeworski, Adam. 2004. "Institutions Matter?" *Government and Opposition* 39(4): 527–40.

Rothstein, Bo. 1998. "Political Institutions: An Overview." In *A New Handbook of Political Science*, ed. Robert E. Goodin and Hans-Dieter Klingemann. Oxford: Oxford University Press, 133–166.

Schwartz, Shalom H., and Lilach Sagiv. 1995. "Identifying Culture-specifics in the Content and Structure of Values." *Journal of Cross-cultural Psychology* 26: 92–116

Scott, W. R. 1995. *Institutions and Organizations*. Thousand Oaks, Calif.: Sage.

Scott, W. R. 2005. "Institutional Theory Contributing to Theoretical Research Program." In *Great Minds in Management: The Process of Theory Development*, ed. Ken G. Smith and Michael A. Hitt. Oxford: Oxford University Press, 460–484.
Sen, Amartya. 1993. "Capability and Well-being." In *The Quality of Life*, ed. Amartya Sen and Martha Nussbaum. Oxford: Oxford University Press, 9–30.
Simon, Herbert. 1997. "Administrative Behavior: A Study of Decision-making Processes". In *Administrative Organizations*. 4th ed. New York: Free Press.
Svensson, Pia 2010. "Värmekameror stoppar vandaler". *Göteborgs Posten* December 31, 2010.
Statens offentliga utredningar. 2009. *En ny kameraövervakningslag*. Report no. 2009:87. Stockholm: Fritzes.
Statens offentliga utredningar. 2002. *Allmän kameraövervakning*. Report no. 2002:110. Stockholm: Fritzes.
Suddaby, Roy. 2010. "Challenges for Institutional Theory." *Journal of Management Inquiry* 19: 14–20.
Tække, Jesper. 2011. "Digital Panopticism and Organizational Power." *Surveillance and Society* 8(4): 441–54.
Weaver, R. K., and B. A. Rockman. 1993. *Do Institutions Matter? Government Capabilities in the United States and Abroad*. Washington, D.C.: Brookings Institution.
Welsh, Brandon C. and Farrington, David P.2007. *Kameraövervakning och brottsprevention. En systematisk forskningsgenomgång*. Brå Rapport 2007:29. Stockholm: Brottsförebyggande rådet.
Wood, David. 2003. "Foucault and Panopticism Revisited." *Surveillance and Society* 1(3): 234–39.
Wood, Neal. 2002. *Reflections on Political Theory: A Voice of Reason from the Past*. 19–27. Palgrave.

Official Documents

Fråga för skriftligt svar 2010/11:448. (Riksdagen: Question for written answer to the Minister of Justice April 15, 2010).
Justitieutskottets betänkande 1997/98:JUU14. (Riksdagen: Committee on Justice, report).
Justitieutskottets betänkande 2010/11:JuU9. (Riksdagen: Committee on Justice, report).Lag (1977: 20) om TV-övervakning. (The TV-Surveillance Act).
Lag (1995:1506), om hemlig kameraövervakning. (Riksdagen: The Secret Surveillance Act).
Lag (1998:150), om allmän kameraövervakning. (Riksdagen: Law on Public Video Surveillance)
Lag (1998:204), personuppgiftslag. (Riksdagen: The Personal Data Act)
Lag (2004:633), om om ändring i lagen (1998:150) om allmän kameraövervakning. (Riksdagen: Law on Amending of the Act(1998:150) on Public Video Surveillance).
Lag (2008:717), om signalspaning i försvarsunderrättelseverksamhet. (Riksdagen: Law on Signals Intelligence in Defence Activities).
Lag (2007:979), Act on the Prevention of Certain Particularly Serious Offenses (om åtgärder för att förhindra vissa särskilt allvarliga brott).
Motion 2010/11:Ju8.(Riksdagen: Documents and Laws).
Motion 2010/11:Ju279. (Riksdagen: Documents and Laws).
Motion 2010/11:Ju304. (Riksdagen: Documents and Laws).
Motion 2011/12:Ju209 . (Riksdagen: Documents and Laws).

Motion 2011/12:Ju261. (Riksdagen: Documents and Laws).
Motion 2011/12:Ju264. (Riksdagen: Documents and Laws).
Motion 2011/12:Ju280. (Riksdagen: Documents and Laws).
Motion 2011/12:M24. (Riksdagen: Documents and Laws).
Motion 2011/12:M852. (Riksdagen: Documents and Laws).
Motion 2011/12:SD66. (Riksdagen: Documents and Laws).
Motion 2010/11: S96005. (Riksdagen: Documents and Laws).

Web Sources

Brå and SKL. 2009. *Kameraövervakning för att förebygga brott—fungerar det?* Accessed July 4 2012 at http://www.skl.se/MediaBinaryLoader.axd?MediaArchive_FileID=77b739f9-592f-4e1e-bfa2-c53c735f9ebd

Datainspektionen. 2010. Datainspektionen inleder omfattande granskning av kameraövervakning. Press release from 24 August 2010. Accessed 4 July 2012 at http://www.datainspektionen.se/press/nyheter/2010/datainspektionen-inleder-omfattande-granskning-av-kameraovervakning.

Grofman, Bernard N. 1997. *Ahrend Lejphart and the 'New Institutionalism'*. CSD Working Paper. Centre for the Study of Democracy, UC Irvine. Accesed 4 July 2012 at http://escholarship.org/uc/item/4s0786k3.

Justitiekanslern. 2011. *Kameraövervakning*. Accessed 4 July 2012 at http://www.jk.se/Arbetsuppgifter/Ovriga-arenden/Kamera.aspx.

Justitiekanslern. 2012. *Kameraövervakning*. Available at http://www.jk.se/sv-SE/Sok.aspx?query=kamera%c3%b6vervakning.

Roberts, Lucy P. 2005. "Video Surveillance Guide: The History of Video Surveillance—From VCR's to Eyes in the Sky." Accessed 4 Jylu 2012 at http://www.video-surveillance-guide.com/history-of-video-surveillance.htm.

SKL. 2010. Viktigt med tydliga regler för kameraövervakning. Press release from 23 February 2010. Accessed 4 July 2012 at http://www.skl.se/press/nyheter_2/nyheter_2010_1/viktigt_med_tydliga_regler_for_kameraovervakning.

SLTF. 2006. *Handbok för kameraövervakning i kollektivtrafiken–ett verktyg för trafikhuvudmän och operatörer*. Stockholm: SLTF. Accessed 4 July 2012 at http://www.svenskkollektivtrafik.se/Global/Fakta%20och%20publikationer/publikationer/Handbok%20f%C3%B6r%20kamerainstallation_061006_slutvers.pdf.

Statens kriminaltekniska laboratorium. 2005. Testa ditt system innan brottslingen gör det: Riktlinjer framtagna av SKL, Polisen, Bankföreningen och Svensk Handel. Linköping: SKL. Accessed 4 July 2012 at http://www.skl.polisen.se/Global/www%20och%20Intrapolis/Informationsmaterial/SKL/Kameraovervakning_riktlinjer.pdf.

Interviews

County Administrative Board of Stockholm. 2008. Interview with the vice director of the unit for permissions and supervision, February 26.

Data Inspection Board. 2008. Interview with a member of the director general's staff, May 29.

Contributors

Fredrika Björklund is associate professor in political science at Södertörn University in Stockholm, Sweden. She coordinates the multi-disciplinary research project Balancing Integrity and Legal Security: A Comparison of Popular Surveillance in Germany, Sweden, and Poland.

Patricia Jonason is assistant professor in the Public Law Department at Södertörn University in Stockholm. Her research interests are principally in privacy issues in the right of access to information.

Elfar Loftsson is professor emeritus of political science at Södertörn University in Stockholm. He has worked extensively on political institutions and environmental politics.

Ola Svenonius is an assistant professor at Södertörn University in Stockholm. He is a member of the Living in Surveillance Societies (LiSS) COST Action IS0807 (www.liss-cost.eu) and member of the international Surveillance Studies Network.

Wojciech Szrubka earned a PhD in political science in 2008 and has worked for several years as a lecturer at Södertörn University in Stockholm. His professional interests include the power of bureaucracies and the question of state power legitimacy.

Eric Töpfer is a senior researcher in political science at the Centre for Technology and Society of the Berlin University of Technology. He coordinated the EU-funded Urbaneye Project on closed-circuit television in Europe, and is member of the international Surveillance Studies Network.

Paweł Waszkiewicz is assistant professor in the faculty of law and administration, University of Warsaw. He has received the Polish Forensic Science Association Award (2008) and the Doctoral Prize from the University of Warsaw (2009). He represents Poland in the Living in Surveillance Societies (LiSS) COST Action IS0807.

Index

A

accountability, 7, 53, 143, 184
activism, 4, 10, 37–38, 47, 58, 75, 83, 95
Act on Public Video Surveillance (Sweden), 32–33, 40, 43, 106–112, 195–196, 203–206
Act on the Protection of Personal Data (Poland), 112
administrative control of, 202–204
advocacy coalition, 13, 172, 182, 184
allmän (see also: public), 11, 32, 41–43, 63–64, 94, 128, 211
Almgren, Birgitta, 77, 80–81, 92
amendment, 32, 83, 168, 175–176, 180–181, 184–185
Anglophone literature, 3, 8–9, 11, 174
anonymity, 158, 163, 171, 185
anti-terrorism acts, 82, 177
assemblage, 4, 6, 14, 63
assembly, 122, 126–127, 173, 176, 180
association, 39, 90, 172–173, 181, 186, 201–202
Aufzeichnung, 99
authorisation, 19, 109
authoritarianism, 7, 9, 47, 58, 72, 81
autonomous, 5, 22, 45, 51, 79, 86, 90, 100, 111, 116

B

Baader-Meinhof Gang, see also RAF, 176
bad-news media, 87
Balance of interests/values, 20, 39, 44–45, 57–59, 104–105, 117, 172, 194–195, 204–205
Basic Law, 51, 103, 112–113, 173, 185
Berliner Abgeordnetenhaus, 28–29, 34, 53–56, 65–66

Berlin Wall, 92
bezpieczeństwo, 36, 66–67, 152, 170
Big Brother, 78
biometrics, 19, 63, 126
biopolitics, 62, 73
black market, 80
bureaucracy, 26, 29–31, 36, 38–39, 43–44, 56, 58–60, 80

C

capitalism, 1, 80
Closed-Circuit Television (CCTV), 1, 4, 6–8, 102, 109, 112, 155–157, 176–184
Christlich Demokratisch Union (CDU), 28–29, 53, 55–56, 179–180
citizens' rights, 20, 29, 34, 36, 50–56
citizenship, 9, 21, 54
civil disobedience, 185
civil liberties, 5, 9, 75, 186–188
Commissioner for Data Protection and the Freedom of Information (Bundesbeauftragte für den Datenschutz und Informationsfreiheit), 29, 53, 114
Communism, 1–2, 6–8, 12, 20, 25, 47, 51, 85, 87, 131
computerisation, 71–75, 90
consumerism, 55
convergence of legislation (see also Europeanisation), 118–120
corruption, 80, 140–141, 177
Council of Europe, 101, 108, 122–123, 126–127
county administrative board (Sweden), 33, 61, 197, 202–203
crime prevention (see also deterrence), 9, 27, 39–43, 46–48, 54, 56–57, 84–85, 121, 132, 142–145, 149,

216 Index

153–155, 159, 169, 179, 182, 197, 202
criminal activity, 25, 41, 142, 197
criminal hot spots, 182–183
criminal justice, 143, 153
criminal victimisation, 45, 59, 153–168
criminality, 27, 38, 41, 201, 205, 209
criminology, 4, 8, 56, 84
critics, 19, 29, 36, 39, 44, 47, 49, 60, 69, 83, 173, 182
culture, 4, 71, 81, 87, 148, 153, 168, 183
culture of control, 131–133, 143–145
culture of distrust, 148

D
Data Inspection Board (Datainspektionen), 31–33, 36, 38, 44, 73–74, 110, 199–203, 206, 212
Data Act (Datalag), 73–74, 83–84
database, 35, 173
data retention, 82, 109, 115
data shadow, 173
dataveillance, 75
Datenschutz, 29, 53, 56, 65–67, 96, 114, 125, 186, 188
deliberation, 136
democracy, 1, 7–9, 29, 36, 38, 50, 52, 54, 60, 74, 77, 90, 173, 195
deterrence (see also crime prevention), 42, 142–144, 149, 155–156, 163, 168–169
deutscher Herbst (1977), 75
digital technology, 199
discursive order, 22
discursive practices, 24–26, 30–31, 47, 57, 59, 80
dragnet investigation (Rasterfahndung), 74–76, 85, 175

E
economic liberalisation (during Communism), 80
effectiveness of video surveillance, 2, 6–7, 40, 42–43, 48–49, 54–55, 60, 87, 100, 142–145, 149, 154–155, 157, 159–160, 164, 166, 179, 182, 194, 207
elite, 38, 73, 131, 179
EU Data Protection Directive, 31, 82–83, 98, 101–106
EU Data Retention Directive, 37
European Commission, 121

European Convention on Human Rights (ECHR), 97–98, 101, 106, 108, 118, 123–124
European Court of Human Rights (ECtHR), 101–103, 108, 118–119, 123, 125
European Union, 2, 19, 31, 35, 63, 97, 101, 104, 108, 119–121, 128, 198
Europeanisation of privacy regulation, 97, 106, 108, 115, 118–121
experimental approach, 13, 155, 158

F
facets of privacy, 99, 102
Fascism, 9
fascist, 75
fear habitus, 79
fear of crime, 70, 87–89, 133, 144, 153, 155–156, 182
Federal Constitutional Court (Bundesverfassungsgericht, BVerfG), 34, 52–54, 76, 123–124, 128, 173, 185
Federal Data Protection Act (Bundesdatenschutzgesetz, BDSG), 29, 75, 114–115, 175
Federal Office for the Protection of the Constitution (Bundesamt für Verfassungsschutz, BfV), 185
Federal Parliament of Germany (Bundestag), 29
Federal Police (Bundespolizei), 180
feeling safe and/or secure, 13, 85, 156–157, 159–160, 163–168
Foucault, 1, 4–5, 14, 79, 93, 171, 187, 189, 211
FRA, 82–83, 206
free democratic basic order (freiheitliche demokratische Grundordnung), 75
fundamental right of information privacy (Grundrecht auf informationelle Privatheit), 121

G
gated communities, 49
George Orwell, 1, 78, 173
German data protection (see also Datenschutz), 29, 62, 180
German Democratic Republic (East Germany, GDR), 9, 14, 77–78, 80–81, 85, 91–92
German discourse, 21, 35–36, 39, 46, 50, 52–53, 56–57, 59, 98, 100
glass citizen, 51, 76–77

I

governmentality, 3, 22, 175

individualism, 172, 177
infiltration, 72, 79, 81, 90
informational self-determination, 13, 34, 52–53, 76, 113, 172–173, 184
Innenministerkonferenz (IMK), 54, 76, 85–86, 174, 179
innere Sicherheit (internal security), 9, 54, 75–76, 85, 172, 174, 179
institutionalism, 6, 13, 189–191; norms, 5, 13, 19, 190–194, 200, 207–208; practice, 5–6, 192, 203, 208
integritet (see also privacy), 10, 39, 43, 45, 58–59, 73, 98, 109, 195
internal security, 54–55, 60, 172, 175, 177, 181, 184–185
interpersonal trust, 12, 132–134, 136–137, 148–149

J

judicial-bureaucratic, 26, 29–31, 43

L

law enforcement, 5–6, 12, 75–77, 82, 90–91, 121, 154, 163, 169, 175, 199
liberties, 5, 9, 75
liberty, 4
licensing regime, 83–84, 89
limitation, 122, 162–163
limits, 49, 52, 56, 59–60, 97, 99, 103, 116, 119, 142, 172, 177, 183, 196, 204
lingua securitatis, 80
logic of appropriateness, 134
city marketing through surveillance, 49–50, 133, 169, 181

M

martial law (Poland), 78
mass media, 37–38
maximum surveillance society, 22
media attention, 72
methodology, 4, 10–11, 106, 135, 148, 155, 162–163, 168
Ministry of State Security (Ministerium für Staatssicherheit, StaSi), 77, 79, 91
Moderates, the, 207
modernisation, 20–21, 45–50, 88

municipal, 36, 138, 145–146, 180–181, 200

N

Nazism, 51, 74, 78, 85
neoliberal, 172, 177, 185
new surveillance, 4, 6, 190, 203

O

ombudsmen, 34

P

Panoptykon, 38, 47, 49, 66–67
parliamentary commission, 73–74
parliamentary debate, 26–27, 41–42, 53
parliamentary debates, 40
penal welfarism, 143
Personal Data Act (Swedish), 32, 61, 83, 109–110, 118, 198, 200, 203, 211
Pirate Party, 37
police surveillance, 28, 53, 61, 85, 172, 174, 177
policing: new modes, 71; of risk societies, 174; punitive, 87, 185; repressive, 75; stop-and-search, 175, 183; strategies, 71–72, 75, 85, 174, 176, 183
Polish constitution, 35, 45
Polish data protection, 66, 94, 96: Data Protection Office (GIODO), 35–36, 46, 88–89, 112
Polish discourse, 11, 21, 30, 45–48, 59–60
political practice, 26, 28, 30–31, 58
political science, 3, 6, 8, 11
Political, the, 70–72
political-parliamentary, 25, 29
politicisation of, 22, 26, 28, 76, 83, 85
Polska Zjednoczona Partia Robotnicza (Polish Communist Party 1948–1989), 78
populism, 88
privacy (see also integritet, prywatność, and Privatsphäre), 5, 9–13, 20, 23–24, 26–27, 32–33, 39–41, 43–47, 50–54, 57–61, 69–71, 73, 82–86, 88–90, 92, 95, 97–105, 107–114, 116–118, 120–124, 128, 131, 133, 139, 149, 171–172, 174, 177, 180–181, 184–185, 190–191, 194–197, 201, 204–207, 209

Index

privatisation of security, 9, 55
Privatsphäre (see also privacy), 51
proportionality principle, 36, 44, 46, 56
protector, state as, 60, 131, 181
prywatność (see also privacy), 45–46, 52, 58–60
public order, 97, 175, 179–180
public security, 97, 105, 114, 195
public space, 19, 42, 46, 53, 76, 84, 89, 100, 148, 159, 177–178, 182, 186
public surveillance, 10, 36, 58, 145, 159, 189–190, 195, 197, 208
public transport, 10, 84, 88, 184, 200
public video surveillance, 10, 40, 54, 112–113, 155–160, 163–164, 166, 168–169, 189, 196, 204, 206, 208

R

rational choice, 13, 41, 135–136
Rechtsstaat, 8
Red Army Faction (Rote Armee Fraktion, RAF) (see also Baader-Meinhof Gang), 71, 75–76, 175
regulation of video surveillance, 35, 70–71, 74, 84, 87, 89, 97, 115, 118–119, 201
repression, 76, 78, 80, 88, 92
resistance, 70, 75, 77–79, 90, 173, 185
retribution, 144, 149
risk society, 21, 174

S

safeguards, 27, 32, 97, 102, 106–107
SAP, 71–73
secret police, 71–72, 79, 87
securitisation, 3
security: subjective and objective, 55–57
signs, 41–42, 47–50, 54, 58, 60, 109, 142, 197, 203
social marginalisation, 55
social order, 22–23, 91, 182
social regime, 24, 58
social sorting, 5
social trust, 13, 79
socialist regimes, 72, 79–81, 91
societal polarisation, 70, 76
sociology, 3, 8
Solidarity, 78, 151
SPD, 28–29, 174, 180, 186
state of exception, 174–175
state-citizen relationship, 19–20, 85
storage of personal data, 84, 115, 117, 145, 173–174, 177, 180, 197, 206
streitbare Demokratie, 74

surveillance studies, 3–5, 8–9, 13, 84, 171, 185
Swedish constitution, 120
Swedish data potection, 27, 33, 82–84, 106, 112
Swedish discourse, 21, 36, 38–39, 43–44, 54, 59, 61
Swedish model, 72, 89
Swedish public prosecutor, 45

T

terrorism, 9, 20–21, 55, 57, 85–86, 176
totalitarianism, 20, 22, 50–51, 56, 73, 77, 79, 131
transition states, 9, 87

U

Urbaneye Project, 2, 7, 23
utilitarianism, 44, 58, 135–136

V

vandalism, 113, 147–148, 160
video surveillance: in ambulances, 132, 139–140; cleanliness, 55; camera, 74, 148, 158, 164, 168, 178, 202; of employees, 87, 102–104, 13; in forests, 145–147; funding of, 155, 169; normalisation of, 10, 19, 77, 174–175, 177; open-street, 77, 106, 177–183; outsourced, 20; panoptic, 79, 171, 185, 189; permit for, 32, 73–74,89, 99, 113, 197, 200, 202–204; practices, 1, 6–7, 12, 20, 23–24, 69, 72, 77, 79, 83, 90, 200–201; recording, 46, 48, 96, 99, 102, 107, 109–110, 112–115, 122, 139, 141, 146, 178, 180, 182, 194, 196–197, 199; symbolic, 9–10, 47, 58, 69, 73, 88, 209
victims, 45, 86, 144, 154, 160, 162, 166, 168

W

Warsaw city monitoring centre (Stołeczne Centrum Monitoringu), 145
Warsaw city security department (Zakład Obsługi Systemu Monitoringu), 36, 48–49
welfare institutions, 9, 21, 72–73, 81, 92
Western democratic tradition, 47, 69, 90–91, 153, 189–190
World War II, 80, 82, 143